D0962521

RUSSIANS AMONG US

Also by Gordon Corera

Operation Columba
Cyberspies
The Art of Betrayal
Shopping for Bombs

RUSSIANS AMONG US

Sleeper Cells, Ghost Stories, and the Hunt for Putin's Spies

GORDON CORERA

WILLIAM MORROW
An Imprint of HarperCollins*Publishers*

For information, address HarperCollins Publishers,
195 Broadway, New York, NY 10007.

HarperCollins books may be purchased for educational,
business, or sales promotional use. For information, please email
the Special Markets Department at SPsales@harpercollins.com.

FIRST EDITION

Library of Congress Cataloging-in-Publication Data has been applied for.

ISBN 978-0-06-288941-6

20 21 22 23 24 LSC 10 9 8 7 6 5 4 3 2 1

For Jane

CONTENTS

RUSSIANS AMONG US

Prologue

It was humid enough for haze to rise off the tarmac as fourteen people crossed paths for a few brief moments at Vienna airport on July 9, 2010. The fourteen—all accused of being spies—were changing planes but also exchanging lives.

Ten were going one way. They had been living secretly undercover in America's suburbs, and they were now on their way to Russia.

Among them were a KGB-trained husband and wife from Boston who had stolen the identities of dead Canadian babies and whose own children were now sitting bewildered in Moscow. A New Jersey couple whose grumpy husband had made way for his wife to take the lead in their joint spy venture. Her success in getting close to power had set off alarm bells in Washington. Another pair had moved from Seattle to America's capital to further their spying career. But as with the others, almost every moment of their life in America had been owned by the FBI. The last of the four couples was the oddest: a retired Russian spy and his Peruvian wife. She claimed she had not even known her husband's real name despite decades of marriage.

Then there was a young man who had not stolen anyone's

identity but had fallen into an FBI trap while he was working his way into Washington's circles of power—the trajectory of a new breed of spy. And last, but not least, there was the twenty-something redhead who would gather tabloid attention thanks to a party lifestyle in Manhattan and London and nude pictures splashed over the papers (pictures she had spent the plane ride complaining to the FBI about).

All ten had been betrayed by a man they had known and trusted and who days earlier had made a dramatic escape from Moscow to the West.

Arriving on a plane from Moscow and heading in the other direction were four Russian men. Two of them—bound for America—were still feeling the effects of the beatings they had been subjected to in the previous days. One had helped catch a traitor in the CIA and the agency had been desperate to get him out for years. The other had played a role in catching a traitor in the FBI but his subsequent fate was the cause of regret in the CIA. Two other Russians were heading for Britain. One was a sullen figure, angry at being forced to confess to being a spy when he said he had never been one. He was the source of guilt for Britain's MI6. The last man, a tough former paratrooper, really had been a spy for MI6. Eight years after the Vienna swap, his former colleagues in Russian military intelligence would smear a deadly nerve agent on his front-door handle, spiraling relations between Russia and the West into an even darker place than anyone would have imagined that sunny July day.

WATCHING THE TWO groups closely was a small group of intelligence officers from the West and Russia. Many had spent their entire professional careers battling each other in the shadows. Now they were just yards apart. For years, even within their own intelligence bureaucracies they had been regarded as dinosaurs—

ageing prizefighters still throwing punches at each other in the ring even though the crowd had long departed. One of the Russians in particular had devoted much of the past quarter of a century to entangling his adversaries in a web of deceit. His American adversaries thought that at long last they had the better of him. In Vienna, one side seemed to have won, the other to have lost. But that only made sense if you thought this was the end. It was not.

That evening Vice President Joe Biden appeared on *The Tonight Show* on American TV. The spy swap was the talk of the town. "Do we have any spies that hot?" Jay Leno asked Biden, referring, inevitably, to the redheaded woman sent back to Moscow. "Let me be clear—it wasn't my idea to send her back," Biden said to laughter. That was true. He was one of those who had opposed the plan to arrest the Russians and engineer a swap but had been overruled after a heated debate in the White House Situation Room. His comments fit in with a deliberate strategy from Washington to play down the significance of what had just taken place in Vienna—to treat it as an inconsequential event. And for the world watching it all seemed like some kind of bizarre retro-throwback, a hangover from the past, a last hurrah of people who could not quite let go of the Cold War. That was a mistake.

Introduction

THIS IS A book about ghosts. The ghosts of spies past have haunted relations between Russia and the West even as the Cold War ended. The Cold War was fought through espionage and defined by it in the public mind. But when that conflict suddenly ended, the spying did not stop. Repeated cycles of treachery and the hunt for those responsible were an obsession for a small band of spies and spy-catchers on both sides. Neither could let go. And this obsession mattered, since the spy wars have continued to shape relations between the two sides over the decades, playing their role in the rise of Vladimir Putin and his drive for revenge.

Ghost Stories was also the code name of the decade-long FBI investigation into Russians living under deep cover as sleepers in America. It was a fitting title, since these were people who had been resurrected from the dead in graveyards as part of Russia's "illegals" program. The story of these spies and those who pursued them is told here for the first time in detail but set against the broader story of espionage between Russia and the West. A confession: I was one of those reporting on the events surrounding the Vienna swap who thought it was all a bit peculiar. Hindsight is a wonderful thing, and this book sets out to explain why those events were not just the last echoes of the past but also foreshadowed a darker future. The scene at the Vienna airport offered a snapshot of a normally hidden Russian intelligence program, and the blinding flash of publicity illuminated both the tail end of a program running from deep in the Cold War and the beginnings of a new Russian strategy that would replace it.

One veteran of the CIA shakes his head in awe as he ponders how his old adversary has the mind-set to send people to live as "sleepers" in another country for decades, patiently burrowing into the heart of their target, waiting for years to act. This was evidence of the Russians' persistence and patience in targeting its adversary—qualities that have not always been appreciated. Meanwhile, on the Western side, the focus on Russia from the end of the Cold War onward has been hazy and erratic.

As a result, the illegals arrested in 2010 were portrayed as something of an oddity and certainly not dangerous. "They successfully infiltrated neighborhoods, cocktail parties and the PTA," one of their lawyers said, mocking the charges his client faced. This was a narrative deliberately reinforced at the time by the US administration, which was in the middle of an effort to reset relations with Russia. Seeing Russian spies either as figures of fun or as all-powerful is a mistake. The reality was that the illegals were involved in building networks and putting down roots that could have resulted in long-term damage. A previous generation of Russian illegals in the 1930s and 1940s had played a key role in helping steal American atomic secrets and eating away at the heart of British intelligence from the inside. There was real risk from their work and their mission reveals much about how Moscow sees—and sometimes misunderstands—the West.

But they were also people. There are complex personal stories at the heart of this tale. Imagine being a child brought up in suburban America, pledging allegiance at school and running lemonade stalls for your neighbors, but then coming home one day from a pool party to find the FBI all over your house. And then being informed that your parents were not Americans but Russians. And then two weeks later being on a plane to Moscow. It is no wonder that the work of the illegals became the inspiration for *The Americans,* a TV drama set in the eighties. And the real-life spies are sometimes more extraordinary than those of fiction.

The ghost-stories roundup of 2010 illustrates the pivot in Russian espionage that was taking place and changes that were not appreciated at the time. "The Western world can't bring itself to believe to what extent it is transparent and vulnerable to Russian illegal intelligence," a person who worked inside Russia's top secret illegals directorate argues. Moscow has long sought to exploit that openness but the way it does so has changed. The end of the Cold War did not end the illegals program, but the new interconnectedness of the 1990s and then the post-9/11 era have changed the way it goes about it. That led to a new breed of spy, Anna Chapman being one example. This is not our parents' Cold War.

Vladimir Putin is a practitioner of the martial art of judo. It allows a weaker opponent to defeat a stronger adversary. Rather than confront that opponent head-on, the trick is to leverage their strength to throw them off balance. That is what Russia has done with the West. Using the West's openness was at the heart of the illegals program, but the Kremlin's judo has evolved since 2010 as the internet and social media have offered new opportunities. The internet's fundamental features make it a perfect place for those who want to obscure or hide their identity—opening the way for what I call the new "cyber illegals."

One reason the 2010 illegals story was downplayed was that the Russian spies did not get hold of classified information. But this was another mistake. What if spies are not after secrets but influence? The Kremlin's agents have learned to marshal espionage, influence operations, and use technology in a novel way as they engage in a conflict with the West that for many years went unrecognized.

These changes were slow to be appreciated in the West, partly a result of the deliberate playing down of the events in 2010 and partly because of the continued hold of a Cold War mind-set. This is a story that matters not just because of what it tells us about Russian and Western espionage but also because of its consequences and repercussions. The pain and humiliation of the defeat in 2010 would not be forgotten in Moscow. It left scars on Putin personally. Russia's leader, who worked with the illegals program as a young KGB officer, has risen to power and cemented his hold on it thanks to a story he told his people about spies and treachery. Revenge for that humiliation would come served in a bottle of perfume eight years later and in a new campaign unleashed against the West by the Kremlin. Russia is now barely out of the news. But that also has risks.

THE CIA'S APPRENTICE spies who have been chosen to operate in Moscow are given an extra level of training on how to spot the surveillance teams that will be following them. Identifying the tiny telltale signs that suggest the woman in the shop or jogger on the street is not just a member of the public but actually a Russian surveillance operative takes an extra degree of preparation required from anywhere else. After a training mission out on the streets, the students report back to their instructors on which of the people they thought were really watchers. But some trainees become so hyperaware they get it wrong, believing that ordinary people are in fact operatives trailing them. These false positives are known as "ghosts." A form of paranoia is always a risk for those who work in intelligence. It can also apply to nations as well. Russia and the West have sometimes seen "ghosts"—the hidden hand of the other side when it is not there, especially now.

Trying to distinguish between the apparition and the real is difficult but a clearheaded appreciation is vital to avoid the dangers of miscalculation.

The aim of this book is not to demonize Russian spies, let alone all Russians. It is a mistake to not try to understand how the world looks from inside the Kremlin. A mind-set that Russia is a besieged fortress under subversive attack from the West—and particularly the CIA and MI6—has played an important role in justifying Moscow's actions. Spying is supposed to illuminate the other side's intentions, but it can also increase tension and drive conflict. This book is based on interviews with dozens of intelligence officers in London, Moscow, Washington, and elsewhere, including some who have served as illegals. Many have worked at the front lines and in most cases they spoke anonymously, but their accounts help explain how we got to where we are today.

It is tempting to talk about a "new Cold War." That conflict is long gone. There is a new one that is being fought today with both old techniques, like illegals, and new ones. And the best place to start this story is at the moment that the last conflict ended.

1

Three Days in August

August 18, 1991

THE SMALL BAND of Western spies operating in Moscow had learned to trust their instincts. And that Sunday afternoon some of them could sense something strange in the air. For CIA and MI6 officers working undercover in the capital of the Soviet Union, the shadow of the all-powerful KGB was everywhere. It was on the streets and in their bedrooms, black cars trailing them, eyes watching them, microphones listening in. They had received the most intense training that their agencies offered in order to survive in the belly of the beast—the bleak and unforgiving capital of their adversary. CIA officers would walk the streets continually so they knew every crossroads and alley better than those in their hometowns. That was so that when the moment came to "go black" and lose a tail they might have a shot. The phrase "Moscow Rules" referred to this highest level of tradecraft—or spy skills—required to contact Russian agents who were providing secrets to you. There was no room for error, since one mistake might mean leading your KGB tail to them. You would be expelled. They could end up dead.

The MI6 station was smaller than its American counterpart—only a handful of officers. The head of the station had been a special forces officer and he liked to run. One benefit was that it let him stretch his surveillance team. Perhaps you might be able to lose them for a second or two in order to drop a package off or pick it up. In order to keep pace, the KGB assigned an Olympic runner to stay on his heels. That worked until the MI6 man, running on the outskirts of town, decided to jump into the Moscow River and swim across just to annoy his surveillance team. But late that Sunday afternoon there was a mystery. When the youngest member of the MI6 station went for his regular run in a nearby park, he could see the usual surveillance was there as he began. But then, as he became increasingly breathless in the summer heat, he realized the surveillance had vanished. No watchers. No parked car. That had never happened before. Catching foreign spies was the KGB's priority. What could have been so important that they were called away?

The mystery was solved the next morning when the spies turned on the radio. Hard-liners were seizing power in a coup. The country's leader, Mikhail Gorbachev, who had been progressively opening the country up, had been detained in the Crimea. Some of the KGB's best surveillance teams—normally reserved to follow foreign spies—must have been pulled back to deal with events. The KGB was the sword and the shield of the Soviet state. The bulk of its half-million personnel were devoted to internal security; its mind-set was the constant search for enemies who threatened the grip of the Communist Party. The chairman of the KGB, Vladimir Kryuchkov, feared Gorbachev's reforms were undermining the pillars on which the Soviet Union had been built, leaving it at risk of collapse, and he was one of the plotters. That Monday morning, columns of tanks made their way across the capital, scattered amid the usual morning traffic as people returned from the weekend. But there was something tentative about the move; the soldiers seemed unsure how to act. The CIA's chief of station, David Rolph, had only arrived three months earlier from Berlin ("at least I'm going somewhere stable," he told colleagues) and now found himself in the middle of a crisis no one had predicted. He told his officers to fan out across the city and find out what was going on.

In the place known as "the Woods" or, more formally, as "Moscow Center," elite KGB officers watched with surprise from the window as the endless column of tanks trundled toward the city center. Yasenevo is set among trees half a mile from Moscow's outer ring road. From 1972 it has been home to spies whose job it is to steal secrets around the world—then known as the KGB's First Chief Directorate. Past the barbed wire and guard dogs sits a gray 22-story building set in a park by a lake. Most of the staff was shuttled in every morning in anonymous buses but the most senior generals were housed on-site in a compound of dachas.

At the moment the coup began, the man in charge of the First Chief Directorate had been playing tennis at Yasenevo. That was because Leonid Shebarshin was not one of the plotters. A few weeks earlier, Shebarshin had stood next to KGB chairman Kryuchkov as they addressed the new political leaders—men like Boris Yeltsin, who had just been elected president of Russia by popular vote. Russia was still part of the Soviet Union. But the control that the center exerted over its empire was crumbling as republics—including Russia itself—were flexing against the Communist Party that ruled over them. Shebarshin sensed that day that the politicians were more interested in their own power struggles than the warning he was there to deliver about the "main enemy"—the United States. "What was not changed was the ambition of the US to weaken the Soviet Union and deprive it of the status of a great power," he told his restless audience. Washington's "new world order" meant American dominance. The Cold War had not really been an ideological struggle, he believed.

It was really about Western resistance to the idea of Russia as a great power. Shebarshin's worldview had surprisingly little to do with communism. As that faded, what remained was a resilient core of nationalism. The Soviet Union covered nearly a sixth of the world's surface. But size was also weakness. Western spies had been supporting separatist forces trying to fray the bonds of the USSR, Shebarshin explained. He was frustrated that no one seemed to be listening. But he had not realized how far the man next to him would go. A few weeks later, the KGB chairman and other hard-liners feared a treaty that was about to be signed would break up the USSR and so launched their coup. Now, from his spacious office in Yasenevo, Shebarshin watched on Monday morning as the tanks rolled into the city. But they stopped at red traffic lights. That hardly looked ruthless. The spymaster was hanging back—not committing either way. Ironically, the man in charge of the KGB's international espionage operations then turned on CNN to find out what was happening in his own country.

Thousands of miles away on that same Monday in August, there was proof that Shebarshin's First Chief Directorate was still in the game. An American intelligence officer went for a walk to a park only a few miles from the headquarters of the CIA in Virginia. He liked Mondays as he knew there were fewer FBI surveillance teams operating. Underneath a footbridge he left a package containing highly classified information about intelligence operations the United States was mounting against the Soviet Union. He retrieved another package containing twenty thousand dollars in cash and a message. He was one of two spies the KGB had deep inside its opponent's intelligence services. This spy had been active for more than a decade but not even the KGB knew his real name. And the pair were not the only spies the KGB had in America—they also had their own officers, living as sleepers.

• • •

HIGH IN YASENEVO'S tower, those who controlled the KGB's most prized spies had also been watching the tanks. Floors fourteen to twenty on the main tower were home to Directorate S. Its work was kept secret even from other colleagues in the building because this was where the KGB ran deep-cover agents who lived abroad under false identities—illegals. Traditional spies work under diplomatic cover in a foreign country—posing as something like a second counselor for trade. If such a spy is caught, they have diplomatic immunity and can only be expelled. Other spies work under nondiplomatic cover, as, say, a businessman. They have no diplomatic immunity. In the Russian terminology, they are "illegal." Many countries undertake this kind of spying but Directorate S takes things a step further. A deep-cover Russian illegal can be not just operating under cover of a different occupation but can take on an entirely different nationality. The Russian will not—to all appearances—be Russian but instead be German or Canadian or even American. They can spend decades undercover in a different country, burrowing deep into their target society—sleepers. Some will live and die in a foreign land, buried in a graveyard under a name that was never truly their own. Illegals are the pride of Soviet and then Russian intelligence—having assets deep inside enemy territory provided a sense of power and reassurance and an edge over their adversaries. Were they worth the vast investment? "My experience tells me that this practise quite justified itself," Leonid Shebarshin said.

When the 1991 coup began, a legendary figure made a surprise reappearance in the corridors of Directorate S. Yuri Drozdov had retired a few months earlier as the head of the directorate. But in August as the tanks rolled, he returned to the duty room, with a sparkle in his eye. "We'll restore order in the country! We'll clear up the mess. It's about time!" he told surprised staff. Lean, with a long face, Drozdov had entered Berlin with the Red Army at the end of World War II and by 1962 was a KGB illegal in Germany. Drozdov escorted American lawyer James Donovan across East Berlin to organize one of the most famous spy swaps of the Cold War, when the illegal Rudolf Abel was exchanged for captured American U2 pilot Gary Powers. He then spent four years in New York before in 1979 becoming head of Directorate S, where he ran the illegals for more than a decade. But in his final months in charge, he had found the uncertainty and confusion surrounding the Soviet Union deeply unsettling.

The illegals around the world had become agitated as they watched news of the political divisions and disarray back home. These men and women had dedicated their lives to an idea that communism would transform the world. But what if that was now disintegrating? At an operational meeting before his retirement Drozdov had revealed the depth of their concern. A few illegals, he said, had broken every rule and written directly to him asking what was going on. "Personal letters are being written to me," he told his astonished subordinates. "They ask what is happening in our motherland. They say they can't understand anything! They ask who it is they are working for. Is it for the Soviet Union or for Russia? For the Communists or for the 'new democrats'?" He explained that the illegals feared that the new political leaders might betray them. Some said they would not maintain contact with the center for their own safety until the situation was resolved. They would continue with their long-term missions but avoid short-term scheduled contacts. "I have received not just one letter like this but several," an exasperated Drozdov explained. "What have they been doing? Have they held their own Party congress?" he joked.

The illegals were the glory of the KGB. Drozdov's greatest nightmare was that this intelligence capability—so prized and so patiently built—could be lost. He was sufficiently worried that he began destroying some of the documentation so it could not fall into the wrong hands. Fear may have driven him to back the August coup. But Drozdov's return, like the coup itself, would be short-lived.

Out in the city, the CIA officers sent out by their chief of station began reporting back. They realized their KGB minders were absent and so for the first time, they took the risk of meeting contacts quickly on street corners without the usual careful preparation. They visited the airport and TV stations—all the places you would seize first in a well-planned coup. But no one had secured them. The whole thing was starting to look like something a group of desperate men had cooked up quickly after too many vodkas rather than a well-oiled operation. Crowds of ordinary people were taking to the streets as a light drizzle fell. Barricades were going up. The plotters had failed to arrest Boris Yeltsin and he would become the focal point of opposition. After decades of the tightest control, everything seemed to be unraveling at bewildering speed. The CIA officers could see the army waver. The coup was a last, desperate attempt to prevent the demise of the communist system. Instead the plotters had hastened its death. Gorbachev flew back to Moscow to reassert control. The plotters were arrested. But the chaos did not end as the state seemed to unravel.

Shebarshin was summoned to see Gorbachev. He was offered the temporary chairmanship of the KGB. As he walked out of the meeting, he could hear protesters outside. He knew what a revolution sounded like from his time in Iran in 1979. He took an underground tunnel to the Lubyanka in central Moscow—the imposing headquarters of the KGB. It was eerily quiet inside. As the light faded, Shebarshin could see Dzerzhinsky Square out of the window. After the 1917 revolution Felix Dzerzhinsky had led the Cheka. Its mission had been to use whatever means necessary to preserve the hold of the Communist Party against domestic and foreign counterrevolutionary forces. Dzerzhinsky led a "Red Terror" in which countless were killed and yet the KGB had built a cult of personality around him. To this day his successors—who portray themselves as guardians of the state battling against subversion—are known as "Chekists."

That August evening protesters gathered around the ten-ton statue of Dzerzhinsky in the square. For them it was the symbol of oppression. There were excited speeches and slogans, an air of the impossible suddenly becoming possible. The crowds maneuvered a crane into place. "Death to the KGB," some shouted. There was at least one undercover CIA officer among the crowd. Despite the suspicions of the KGB that they were manipulating events, the Western spies in Moscow were doing no more than observing, astonished as history unfolded before their eyes. Shebarshin forced himself to watch from a fifth-floor window as Dzerzhinsky was taken firmly by the neck. The man who had overseen countless summary executions remained expressionless as he prepared for his own. His iron legs seemed to give one last shudder and then he toppled. A man in a white shirt stood on the empty pedestal and shook his fist triumphantly in the air. For the crowd it was a moment that signified the end of the old order. The KGB was dead. Wasn't it? In the early hours of the morning, a few KGB officers snuck out of the Lubyanka and left a message on the empty pedestal. "Dear Felix, we are sorry that we couldn't save you. But you will remain with us."

The next day, staff inside KGB headquarters were ordered to seal the doors. Knowledge is power for a spy service, and for the KGB it resided in its files that listed the names of informers and agents at home and abroad. These had to be protected at all costs. When East Germany had seen its revolution in 1989, the offices of its security service, the Stasi, had been overrun. There had been frantic shredding of documents. In Dresden, a young KGB officer on his first foreign posting had watched in fear as the crowds gathered outside his office. He had dreamed of joining the KGB since he was a teenager. He called a Red Army tank unit to ask for protection. He expected them to crush the protests, but they explained they were still awaiting orders. "Moscow is silent," he was informed. He was shocked. It was time to destroy the files. "I personally burned a huge amount of material," the KGB officer later recalled. "We burned so much stuff that the furnace burst." That officer's name was Vladimir Putin. He would never forget what happened when crowds rose up and Moscow was silent.

In August 1991, the KGB in Moscow feared the same fate as the Stasi. Shebarshin opened his safe and pulled out incriminating papers so they could be destroyed. He also took out his service pistol—a Makarov 9-millimeter semiautomatic. He carefully oiled and cleaned it. He gave the order to ship files out to a secure location. Shebarshin was then informed that his time as head of the KGB would be only a single day. Gorbachev returned him to head of the First Chief Directorate and installed a liberal reformer as his superior, with a mission to break up the KGB.

Thousands of miles away, in Langley, Virginia, the cables from their team on the ground were being pored over for every detail. No one in the CIA had seen the coup coming and now their opponent was down. But were they out?

"Their dicks are in the dirt," the head of the CIA's Soviet division, Milt Bearden, used to tell his staff. Bearden was a straight-talking Texan with a swagger to match who was head of the CIA's Soviet and Eastern Europe division. The division was the powerhouse of the agency. Its inner sanctum would become known as "Russia House"—a reference, like so much in the spy world, to a John le Carré novel. But at the moment of triumph, it was a house divided. Many officers had spent their entire career having been working against the Soviet target, but Bearden was an outsider, his last job running operations in Afghanistan. That meant he was viewed with suspicion by the insiders. There were deep divisions over how to deal with the old enemy. Bearden's view was that times had changed and that liaison—sitting down with the other side's spies—offered new opportunities. That view was met with deep resistance from those who thought it was naive to think their adversary was changing. The "insular subculture didn't want to let go of the Cold War," Bearden would later write of his critics; "it had been too much fun." Over the decades to come, and even as it moved away from the center of CIA operations, Russia House would always retain its own unique identity, its work sealed off from everyone else behind walls of secrecy. Its critics would say it was trapped in the past, but inside its walls, its inhabitants believed they were the only ones who understood that, whatever changed on the surface, the opponent they faced off against was patient, persistent, and aggressive, and only they fully appreciated the danger. At the time of the coup they thought Bearden did not get it.

As the coup collapsed, Bearden summoned one of his officers into his spacious office. Mike Sulick was a Bronx-born former marine who had served in Vietnam and had a PhD in Russian studies. He joined the CIA in 1980 and had already done one undercover tour in Moscow. He would eventually rise to be the head of the CIA's clandestine service and a central figure in the intelligence war with Russia, one of the key players in the Vienna spy swap. Sulick that day had his bags packed to head off to Lithuania, one of the Baltic states that had sought independence from the USSR, in order to make contact with their intelligence service. "As I entered Bearden's office, he eased back into his chair, propped his leather cowboy boots up on his oak desk, and broke the news: 'Sorry, trip's off, young man,' " Sulick later wrote. Then Bearden broke into a grin. "But look at it this way. It's not every day the president puts you on hold," he explained. The White House had delayed Sulick's trip while they worked out whether or not to recognize the Baltic states. A few days later Sulick was allowed to travel to meet the Lithuanian spies. Once there, he walked around the local KGB office, a grand building that had been stormed while the coup had been taking place in Moscow. It was littered with documents half burnt and shredded by frantic KGB officers a few days earlier. A portrait of Dzerzhinsky had been slashed with a knife by the protesters who had forced their way in. Sulick went down into the grim, dark cellar in which prisoners had once been tortured in tiny cells. There was even a padded, soundproofed room for those sent mad by their punishment. "The empty cells still seemed faintly to echo the screams of tortured prisoners," Sulick later remembered. He felt a tightness in his chest, an inability to breathe, and he could only stay inside for a few moments. This was the type of memory that stayed with those who battled against Russia in the spy wars and made them determined never to relent.

A few days later Bearden was in Moscow and sat across from the new, reformist head of the KGB. The KGB chief explained he wanted to end the Cold War mind-set. Too much effort and money had been wasted, for example, by putting listening devices into the new US embassy. Bearden slipped the US ambassador a note: "Ask him to give you the blueprints." To the amazement of the Americans, the new KGB chief would soon hand over not just the plans but some of the actual transmitters that had been buried inside the American embassy being built in Moscow. Bearden took it as proof that his new ways of liaison might bear fruit. The KGB chief hoped the Americans would reciprocate with details of their bugging operations on Soviet missions in Washington and New York. He would be disappointed.

The handover of the plans by their new boss stunned the hardened operatives of the KGB. They thought it was madness. "How naïve to believe the fall of the Soviet Union meant foreign intelligence would no longer be needed," thought one officer who had battled America. The KGB's old guard, like those in the CIA's Russia House, were not yet ready to give up the game even if their bosses were sitting down together. And their dicks were not quite in the dirt as much as it looked. One thing the KGB had not revealed was that even as their country fell apart, they had a pair of aces up their sleeve—two spies in the heart of American intelligence. But the reforms were too much for some. Shebarshin resigned soon after. "The Soviet Union is no more but eternal Russia remains. It is weakened and disorganised but the spirit of the Russian people has not been broken," he wrote.

By the end of that year, republics including the Baltics had their independence, Gorbachev was gone, and the Soviet Union was formally dissolved. And so was the KGB, broken up in early December, to reduce the concentration of power. The old First Chief Directorate—the sword—became the SVR, with the task of spying abroad. Even though his statue had been torn down outside, inside the Lubyanka, Dzerzhinsky's picture was still up on the walls, and small statues stood in rooms and corridors like shrines. The domestic security arm of the KGB—the shield—would go through various names in the coming years before eventually becoming the FSB. The names changed and the organizations underwent a crisis of morale in the following years. But what was preserved among a small cadre of KGB officers was a mind-set—the one that Shebarshin had articulated in 1991, in which Russian nationalism supplanted communism and in which spies had a duty to preserve the state to protect the motherland. That would be passed on to a new generation and when Shebarshin, decades later and in ill health, shot himself in his Moscow apartment using his ceremonial pistol, he would be living in a country run by one of his former KGB officers.

The KGB was dead. But it would rise again as something else. Only a few people understood that less had changed than at first sight. They included the small group of Western spies operating in Moscow at the time of the coup. In August 1991, their respite from surveillance had not lasted long. Precisely three days after vanishing in the park on a Sunday afternoon, the young MI6 officer's minders from the KGB were back on his tail as if nothing had happened. The game went on.

On December 26, 1991, a married couple sat in a hotel room in Buffalo. They watched on CNN as the flag of the Soviet Union was lowered for the last time and they wept. The pair, with a newborn son to look after, were far away from home. He used the name Donald Heathfield. She was Ann Foley. The trauma they had felt over the last few months as they watched their country collapse had to be buried deep down and internalized. Ann had developed a nasty skin inflammation. It baffled the doctors since it seemed stress-related in origin. And yet on the outside she gave off the impression of a young woman without a care in the world. The couple had to hide their reaction to events in Moscow because they were KGB sleepers living under deep cover in Canada, pretending to be Canadians. Their long-term target was the KGB's "main adversary." But now the regime they had served and sworn an oath to was gone, along with the KGB that had trained them. Some illegals would use this moment to disappear, discarding their true selves and melting into the West as their adopted selves. But this couple chose to continue with their mission. They told themselves that their country—Russia—still retained their loyalty. "For me, my country is more than just a government or a certain political arrangement. I was serving my country, my Motherland," Ann says. But as they watched the ceremony dissolving the Soviet Union and cried, they felt alone. What was their future now?

2

The Birth of an Illegal

DONALD HEATHFIELD, LIKE his wife, had been born in a cemetery, a ghost rising from the dead. A baby boy had been born on February 4, 1962, in Canada, the third of four children of Howard and Shirley. Six weeks later, on March 23, Shirley found little Donald lying still, a tiny arm sticking out of the side of his crib. Her child had died. Tracey Lee Ann Foley was born on September 14, 1962, in Montreal, the first child of Edward and Pauline Foley. Seven weeks old and just a few days after she had smiled at her mother for the first time, she developed a fever. Within hours, she died of meningitis. As with the Heathfields, the pain of the loss of a child so young never left the family.

But then a quarter of a century later, Heathfield and Foley were suddenly there again, brought back to life by Directorate S.

The twin tragedies had not gone unnoticed. A KGB officer serving in Canada had observed them. He would steal something from these two families who had already lost something irreplaceable—their children's identities. KGB officers had the macabre job of strolling around cemeteries looking at graves for likely candidates, a process known as "tombstoning." The ideal situation was a child who died away from the country in which they were born, with few close relatives, reducing the documentary and witness trail to the death. Once a candidate was found, the next step might be to destroy any documentary evidence of the death. This could be as simple as bribing someone for access to a church registry book and then ripping out the pages. Then came the key—requesting a new birth certificate (a technique that relied on there being no central registry of births and deaths). "It was considered a big success for us when Department 2 managed to obtain children's birth certificates after a whole family died in a traffic or other kind of accident," explains one former member of Directorate S. A birth certificate meant a child could be born again as an illegal.

Directorate S was broken up into departments. Department 2 was the storytellers. Their job was to create a fictional life and to make it plausible enough to stand up to scrutiny from a discerning critic by building "legends" and providing backstories. Officers of the department would draw up paragraphs in two columns. On one side would be the supposed detail of a person's life—Donald Heathfield was Canadian and born in Montreal. On the other side would be the made-up evidence supporting that claim—starting with a birth certificate. If there was a claim that did not have documentation, then there would be a plausible story why. It was painstaking work. If there was any doubt, an entire identity would be discarded. Roughly one in ten attempts would create something considered sustainable against checks by Western security services. Each illegal had a "kurator"—literally a curator of the false identity who would supervise their training and act as a handler once they were in the field.

The Operational Technical section of Department 2 includes a team of highly skilled forgers. What does a French passport issued five years ago look like? What does a Finnish driver's license look like? They study which inks, papers, glues, and even staples are used in target countries so they can be faked or—if blanks can be stolen—doctored with a new identity inserted. A laboratory works on how to replicate the different types of paper and ink and how to artificially age a document in a special oven so a passport can be filled with the backstory of visas and trips and made to look old when it is in fact new. So why not just create entirely fake personas for the spies? A proper check into someone's background would raise too many questions and if fake documents are spotted then it is game over. For long-term penetration, the strong preference was always to get hold of real documents rather than rely on fakes. This meant becoming a "dead double"—stealing an identity of someone deceased and then using it to build a set of genuine documents. That was the route for Heathfield and Foley. They might arrive in Canada and start with a birth certificate. This could be used as the stepping-stone to contact public bodies and obtain other identity documents. Ultimately this would eventually lead you to a real passport, helping create what was called an "iron legend."

So who was the resurrected Donald Heathfield? His real name was Andrey Bezrukov. He was born on August 30, 1960, in Kansk, in remote Siberia, a small town near the route of the Trans-Siberian Railway, home to a MiG fighter base. His parents were often away for work and so he was an independent child, self-contained with a strong inner confidence. Bezrukov traces his family tree back to the Russian conquest of Siberia under Ivan the Terrible in the sixteenth century, when his distant relatives had first come to the region. "For me to forget this is to be left with nothing," he later said. Remembering your roots was important when you were pretending to be someone else. Patriotism would sustain him in his long years far away from home.

In 1978, at the age of eighteen, Bezrukov went to Tomsk State University. His study of history gave him a sense of the uniqueness of Russia's story, a country engaged in what he calls an "endless, painful search for herself between East and West." And it was while a student that he was talent-spotted. Universities are the classic recruiting ground for illegals. Department 3 of Directorate S is in charge of the intense selection process. An ideal candidate is in their early twenties. When a person was younger than that you could not be sure they had what it took to survive. By thirty, they were no longer malleable enough to be shaped into a new person. Spotters looked for those who might have the right set of skills—an aptitude for languages was vital, so was intelligence, patience, adaptability, an ability to cope with stress, and a sense of patriotism. Careful psychological assessments were undertaken. Someone who was volatile or looked like they might drink too much or have too much of an eye for the opposite sex was not suitable. This was all initially done at a distance before a move was made—perhaps, as often in the West, on the recommendation of a professor. Somewhere among the stream of students carrying books to and from class and flirting with each other, the KGB had spotted Bezrukov.

Bezrukov was not recruited alone. The fact that illegals were selected in their twenties posed a problem—relationships. An illegal was destined to spend decades living undercover. It was unrealistic to think they would not engage in relationships. But this posed a danger. If you fell for a local, you would either have to constantly try to hide the truth from them or risk telling them you were not who you said you were. Worse, you might place love over duty and give up your spying career. Anatoli Rudenko, a 1960s illegal, worked in West Germany and London before ending up in the United States as a piano tuner to the rich and powerful—including the governor of New York, Nelson Rockefeller. Rudenko's career ended when he was forced to reveal he had defied orders by secretly marrying a hairdresser in Germany and taking her with him. Plans to use him to penetrate the United Nations and think tanks—including by seducing lonely young women—had to be shelved. It was an example of why human relationships were the key to an illegal prospering or failing.

And so the preference became to send out couples. Marriages were sometimes arranged and manufactured by Directorate S (its officials could even officiate in order to maintain secrecy). An arranged marriage would not just avoid the danger of falling in love but also offer a partner in undercover work. "You would not have to waste your time chasing after girls and risk falling into bed with the wrong one," one illegal was told during the Cold War when a partner was offered to him. "You would not have to explain your absences. . . . [Y]our partner would be a trained agent who could help with communications, photography, drops—with everything. You would not be alone behind the lines." Not all marriages worked. Yelena Borisnovna and Dimitry Olshevsky were sent to Canada under the identity of two dead babies, Laurie and Ian Lambert. Their relationship hit the rocks out in the field. Dimitry moved in with a local woman, while Yelena began to date a British-born doctor. Canadian intelligence arrested them in 1996. The pair were deported, landing on a stormy night in Moscow to be whisked away in a blacked-out van straight from the runway.

Andrey Bezrukov's marriage, though, was no fake. This was an adventure that two young people set out on together. Elena Vavilova was a fellow history student at the university. She had been born in November 1962 in Tomsk, where her parents were academics. She was a cheerful, outgoing child who enjoyed figure skating, ballet, and acting. At university she played the violin as she studied for her degree. There she met Bezrukov. There was a confidence to him and also a sense he might take her out of the well-ordered world she had grown up in. "Andrey offered me something out of the ordinary, an adventure," she later said. The young couple spent the night together in the university library, sneaking in before closing time and staying among the book-lined shelves. But they were caught by the director the next morning. There was a telling off but no punishment. Perhaps it was this spirit of adventure and the willingness to take risks together that got them noticed.

The young couple was approached by a man who had an unusual proposal. Did they want to serve their country? She was only twenty-one. "I wondered what would happen to our relationship," Vavilova thought. "I believe we were selected separately, each of us could have refused the proposal," Vavilova later explained to me. "However, since we were already romantically involved, it was more beneficial to have us as a couple for the training." She would later reflect that love marriages among illegals were better than arranged ones because of the trust that was always there. Their first curator sat down with them and in long conversations began to see if they were suitable—their recruiter talked to them about their backgrounds, their lives, their friendships, and their studies, carefully probing them to make sure they had what it took. The strains of the life they were to live drove others apart but, in their case, it would bring them even closer together.

Why did Elena Vavilova agree to become a spy? "The concept of the Motherland—an amalgamation of everything that is important to you," was her explanation later. Vavilova and Bezrukov were still living in the era of communism when they were approached by the KGB, but it was always as much defending Mother Russia as spreading communism that had motivated them. "For me the main motive that made me agree and accept this job was the desire to prevent another terrible war, like the Great Patriotic War," Vavilova later said, using the Russian name for World War II. "As a teenager, all the films about the war and the suffering the people had to go through and the high price we had to pay for victory, all of this fostered in me a wish to be part of whatever could be done to prevent it from happening again." This was the driving force for many Russian spies from the Cold War generation—the sense of threat to their country and the story of their near defeat at the hands of the Nazis before a victory that came at an enormous cost (one which many feel is rarely acknowledged in the West). Almost every family had lost someone in that brutal war. The illegals' mission was to prevent it happening again by acting as a warning system. The early eighties, when the couple was approached, were years when the need for such warning seemed all too real. In Washington, there was an American president calling the Soviet Union "the evil empire" and who, Moscow feared, might be gearing up for confrontation and perhaps even a first nuclear strike. And at the same time, young men from the Soviet Union, including fellow students from Tomsk, were heading off to fight in a brutal conflict in Afghanistan.

In the West, the word *spy* refers to both heroes—the James Bonds—and villains—traitors like the Kim Philbys. But in Russia they separate the two concepts with different words—a spy is a betrayer of secrets. Meanwhile, their word for heroic intelligence officers translates more closely in English to "scouts"—in the sense of someone who is working behind enemy lines to scout ahead and report back, providing advance warning. The separating of the two ideas makes it easier in Russia to lionize the heroes and demonize the villains. It also explains how the illegals saw themselves as operating behind enemy lines in order to protect the motherland. They were the "soldiers of the invisible front."

Soon after, the couple was married. Bezrukov's parents came to the modest reception in Elena's parents' house. A few years later they would be married for a second time, this time as Donald Heathfield and Ann Foley (she chose to use her new middle name rather than Tracey). "The gap between our weddings was short, a few years, but those years were very intense," Vavilova later explained. There was no honeymoon the first time. Instead, Bezrukov and Vavilova simply vanished from the life of friends and family as they headed to Moscow. Despite the excitement of being in the nation's capital and the luxury of a two-bedroom apartment, this period was challenging. "The years before we left were quite taxing physically, psychologically and intellectually," she would say. By the end of every week, they would be exhausted. They had to learn all the traditional "tradecraft" involved in being a spy—out on the Moscow streets they practiced brush-pass contacts, where one person hands over items to another surreptitiously—and tested on whether they could spot surveillance. Inside, they learned martial arts, how to shoot a gun, and how to evade a polygraph lie-detector test. But there was much more to the creation of an illegal than attending regular spy school. For a start, the training was provided to them as a couple by a small group of tutors. They did not mix with other recruits to protect their identity. But it also involved going much deeper. Each illegal required a staggering investment—around four years and by some estimates a million dollars. This compares to CIA and MI6 training for new officers, which is measured in months, not years. "It was not a mass production," one head of the KGB said. "You do not train illegals . . . in the classes. It's a piecemeal operation. You work with an individual, one on one. And only in such a way, we can make them look like an Englishman or a Spaniard or a German."

As the Moscow snows came and went, Bezrukov and Vavilova's education involved former illegals educating them in the history of the elite spies whose ranks they were to join. Moscow had developed its specialty in illegals after the 1917 revolution. Many countries did not have diplomatic relations with the communist Soviet Union so diplomatic cover was not an option. In those early years, there had also been a pool of ideologically committed communists of various nationalities who were willing to spy for Moscow. This led to the heyday of the illegals in the 1930s and 1940s. Richard Sorge operated undercover in Japan, moving in the highest diplomatic circles to provide vital intelligence about Tokyo's relationship with Nazi Germany. In Europe, illegals recruited people who were students and some would slowly work their way into positions of power and influence. Some, like Kim Philby and the Cambridge spies in the United Kingdom, would reach the highest echelons of Western intelligence agencies. In America, illegals worked with the atomic spies who stole the most sensitive secrets imaginable, which proved vital in allowing the Soviet Union to avoid defeat in the early Cold War. Many of these early international illegals were rewarded with execution in Stalin's purges.

In the mid-1950s, the KGB began a push for a new generation of homegrown illegals. They never quite matched their predecessors, although there were some successes. Among the best known was Konon Molody, who turned into Canadian Gordon Lonsdale. He came to London and ran a spy ring stealing naval secrets. Molody had previously worked as an understudy in the United States for another of the great illegals—Rudolf Abel, a remarkably talented individual who had been born in Britain and served in World War II on the front lines before embedding himself in New York. These were the footsteps Bezrukov and Vavilova were to follow in.

The illegals were treated as heroes in the Soviet Union, far more than their spy counterparts in the United States or Britain. There were stories of their daring undercover operations in World War II. The most fabled was Nikolai Kuznetsov. Films and books were based on his time posing as a member of the German occupying army in Ukraine during which he killed six senior Nazis. The accounts of World War II illegals were a central part of Soviet popular culture promoted by the KGB. One fictional work was turned into a 1973 TV series called *Seventeen Moments of Spring*. It featured a deep-cover illegal, a Soviet version of James Bond, who posed as a German aristocrat to infiltrate the Nazi SS and who prevented the Nazis from negotiating a peace deal with America. The series was a massive hit, reaching up to 80 million viewers, and was constantly repeated, embedding itself in the popular mind. Vladimir Putin was twenty-one when the series was first shown and he became desperate to sign up for the KGB. One of his first jobs in East Germany in the 1980s, he would later boast, was to work with illegals.

The CIA and MI6 have never had quite the same capability. They do use what Britain calls natural cover and the Americans call non-official cover (NOC) for their spies. When someone is recruited as a CIA NOC (pronounced "knock") every trace of their contact with the agency is obliterated. Sometimes these officers became annoyed because they had to give up their salary for the cover job, which exceeded their CIA pay even though they felt they had earned it by doing both jobs. They also often felt on the margins of the CIA. Occasionally an individual with a private sector background suddenly gets a surprisingly high-level job at Langley—the reason is they may have spent their career as a NOC. The closest the United States got to Russian "illegals" was the recruitment of dual nationals into its NOC program. So a French-American recruit might end up as a French businesswoman but actually be a CIA officer. But the United States and United Kingdom have never really repaid the favor of sending their own long-term agents into Russia posing as citizens of a third country for extended periods. Why? There is a simple answer. What CIA or MI6 officer at the start of their career would relish spending two or three decades in Russia? The advantage for Russia was always precisely its weakness during the Cold War and after. People from other countries—including undercover Russian spies—want to live in the West. People in the West are less likely to want to go the other way and spend two decades working in Volgograd pretending to be a Ukrainian. It also takes patience, sacrifice, and long-term thinking to create an illegal, something the KGB's competitors were not always so good at.

What were illegals for? The simplest answer is that they were there to do the things that other spies could not. Konon Molody was used to run spies where it was feared that KGB officers under diplomatic cover might be spotted. Other illegals were used to gather specific types of intelligence—for instance, by being trained scientists who could infiltrate biological research establishments or analyze technical data from agents there. Others were deployed purely in case of war. If conflict did break out, then diplomatic relations would be cut and these illegals would take over the running of agents from embassy spies. Some sleepers—also called *konservy*, or "preserves"—were there purely to carry out acts of sabotage in the event of war. The scale of Soviet and later Russian investment can only be understood when you realize that some illegals were trained for years and then spent decades in their target countries and yet were never called on to do anything operationally. They were a sign that Moscow played the long game when it came to spying. But other illegals—like Bezrukov and Vavilova—were not "sleepers" in the sense that they were dormant, waiting to be activated. Rather, from the beginning of their deployment they were to be engaged in starting to work their way into Western society with the purpose of gathering intelligence.

An illegal could get deep into their opponent's society in a way that a legal spy could not. That allowed you to understand your adversary and also meet people who would not engage with a Russian. Doing this required an actor's talent, Bezrukov would later say. But an actor turns on his character for a performance and then returns to his normal self. That is not an option for an illegal. You had to become someone else—and never let that mask slip—and yet never lose your real self. Isolation was a constant worry. The illegal trainees were carefully observed, including by psychologists, to see if they would crack under the stress or if they could hold their cover. This was different from worrying if a secret agent would break under interrogation and torture. The stress instead would come from decades living in a foreign land. One person described the challenge as similar to training a flight crew of cosmonauts who were going to go out to space. The bonds between those out on a mission and those supervising them back home needed to be strong. A traitor among such a tight circle was almost unthinkable because of what it would mean. What Bezrukov could not have known was that at the very moment he was being trained in Moscow, the man whose fate would eventually determine his was undergoing his own training. A veteran of the Afghan war, Alexander Poteyev had been selected to attend the Red Banner Institute in Moscow—preparation to become a KGB intelligence officer.

Learning the language of their target countries took up the majority of Bezrukov and Vavilova's day. They had teachers who had lived in the West but were also given videocassettes of films whose dialogue they would copy and learn. They had to push their own language back in their minds so that the first word they reached for was in their new tongues of English and French, even if they swore. It was not just language but lifestyle. You had to wear the clothes, eat the food, smoke the cigarettes, and even use the razor blades of the target country. "We trained authentic Americans and Englishmen on Soviet territory," explained one head of the KGB. "Habits of how to fill out forms in a London post office; how to pay for an apartment in New York." You had to listen to the radio, watch the TV, and read the newspapers while instructors—often defectors—would test you. When one illegal returned to Moscow, his boss spotted him at the airport in a restaurant, slowly eating his dinner with a knife and fork like a "prim Englishman"—even smoking a pipe as if he were back in England. Vavilova had to learn tiny details that might give you away—for instance, in America when you counted with your fingers you did not bend them like you did in Russia. She would also have to learn what she called the "peculiar American optimism," which meant "keeping your face smiling." No more dour Russian looks for her.

There is one fear that haunts Moscow Center. Could those undercover Russians take to their new lives a little too much and literally go native? During training, illegals were tested with dummy missions. They would be tasked with meeting an agent, but the real purpose was for the agent to report back on the prospective illegal. In some cases illegals were even given a truth drug. Bezrukov's final test involved being given an attaché case with a false bottom containing, he assumed, drugs to hand over to someone he was told was a crime boss. It was supposed to be a straight handoff, but when he met the recipient, they forced him into a car. The case was empty and Bezrukov was subject to interrogation—with a gun pointed at him—to see if he could cope with the unexpected.

When the training was complete, Andrey Bezrukov was going to disappear. Before they were posted abroad in the late eighties, Bezrukov and Vavilov, now about to become Heathfield and

Foley, like every illegal, had to swear an oath to the party, the homeland, and the Soviet people.

There was always trepidation, their spymasters would recall, as an illegal was finally dispatched. It was like sending a child out into the world. Drozdov was something of a father figure to the illegals in these years and would personally check on their progress. But now it was time to let them go. Another head of the directorate compared the moment to taking someone you had just taught to swim and sending them far out to sea. You did not know if they would have the strength to cover the long distance that lay in front of them.

As the moment arrived, Bezrukov and Vavilova sat in front of an empty suitcase in their Moscow apartment. All the clothes they would travel in had to be carefully purchased outside of Russia so that nothing could give them away. They had cleared out every pocket so there was nothing like a coin or ticket stub. Into a box they placed the mementos of their old life that would be left behind—their love letters, Communist Party membership cards, and even their weddings rings. These would be handed over to the KGB for safekeeping. But there was one thing that Elena Vavilova held in her hand that she could not let go of—some pictures of her childhood. They were her last connection to Tomsk and to the life she was saying good-bye to. Those would come with her. It was a mistake that would come back to haunt her.

What was the most difficult emotional experience in departing for a new life? The farewells to parents were hard (there was a cover story of going to Australia). But the real fear was never returning. "There was a possibility we would never come back. We even contemplated the possibility of dying there," Vavilova later said. "Strangers in a strange land, under alias." Bezrukov was haunted by the thought of the Canadian cemetery his false identity had been born in and the fear he might be buried far away from Mother Russia. The tombstone would simply read "Donald Heathfield." If you were an illegal, even your death would be a lie. That was the life the young couple had committed to as they set out.

3

Strangers in a Strange land

THE MONUMENT WAS at a picturesque site in Canada, although neither the man nor the woman who met each other there will say precisely where it was. The pair were playing at being tourists and strangers. The woman walked down the steps. She stopped for a moment looking for something in her handbag. The man happened to be standing in front of her. In these situations, a camera could be your best weapon. "Good morning. May I take your picture? You look so good in the sunlight," the man said. Yes, she replied, and they began to talk. It seemed like a chance encounter but, in reality, it was the opening scene in a movie whose script had already been written. As their legend would have it, this was where the pair's romance began. But the truth was they were not strangers. Rather they were already a married couple. The encounter at the monument was the cover story for where and how Donald Heathfield and Ann Foley first met and how their relationship began.

THE PAIR HAD arrived separately in Canada in 1987, the Cold War still under way. There was excitement at the chance to prove themselves but also fear. The initial journey to a target country was a moment of high risk. There had to be no chance you could be traced back to Russia. So a journey might go first to Eastern Europe, and then to Cyprus, to the Middle East, to Asia, and finally to Canada, at each stage a different set of documents used and then discarded. For Ann Foley, the final entry into Canada was the moment of greatest fear. "You also have to keep your emotions in check, keep calm, not show you are flustered or afraid," she later recalled. There should be no sudden movements or looking around. But she had nothing to worry about. The Canadian authorities still do not know what identity the pair arrived under or even the date they came into the country. Once the disposable identities had been tossed away, the illegals switched to their new settled identities. First they had to meet and then melt into their target society. Canada was a long-established stepping-off point (the "host" country in the center's terminology) to prepare to reach the "target country." At least four of the eleven ghosts who would be the target of the 2010 arrests would have some kind of Canadian documentation. Canada was the ideal launching pad for illegals into America. The culture and language allowed an illegal to acclimatize and build up their identity while border and document checks were largely ineffective. "Canada is a lot like the US, only colder and with fewer people," a KGB officer explained to one illegal in the 1970s.

Heathfield and Foley's mission was long-term penetration of the "main enemy." But what is staggering is that they would spend more than a decade building up their cover before they actually went to live in the United States. That was how long Moscow Center was willing to wait. There was occasionally contact with Moscow Center as orders and money were sent, but their main job was to forget Russia and immerse themselves in Canada. Vavilova would observe other young women whom she saw on the street or met and then try to copy their gestures or their style of conversation. A job was vital partly as it started you on a career that would lead to contacts but also because you needed to explain where your money came from. Some illegals started a business (with money from Moscow); Heathfield had little help though. "I had to get an education again, look for work, create a business . . . without anyone's help and with minimal resources," he later said. In Montreal, Foley enrolled in a course at the Computer Institute of Canada and took a job in accounting at a garment factory. Heathfield worked in accounts at a Honda dealership. It just about covered the bills. They were tough years with long hours, the hard graft of being an illegal. They moved to Toronto and on June 27, 1990, they had their first son, Timothy.

"Every undercover agents' family have to decide, whether to have children at all," Foley would later say. "This is a difficult decision to take." You were bringing a child into the world whose family was living a lie. This was a heavy responsibility and some illegals decided against it. "We carefully weighed this, of course, discussed a lot," she later said, acknowledging that "our leaders also had concerns." But it was something the couple had always wanted. They also knew that from the outside, they would look more "normal" if they had children. Even the act of childbirth has risks. In the famous drama about illegals, *Seventeen Moments of Spring,* an illegal gives away her identity by crying out in pain while giving birth. The problem was she had done it in Russian. When it came to her time, Vavilova as Ann Foley took extensive precautions, attending prenatal classes to learn how best to control herself. She refused anesthetics to keep a clear mind and made sure her husband was present in case anything went wrong.

The summer after the birth of their first child, the couple watched the coup in Moscow and then the collapse of the Soviet Union. The regime they had sworn an oath to was gone. Suddenly, they were on their own. But where some illegals may have given up, Heathfield and Foley did not. It was a "painful period," Heathfield later acknowledged. "We could not receive support from the Center. We had to fend for our ourselves and cover all our expenses," recalled Foley. A sense of patriotism endured even as the ideology they had sworn an oath to vanished. But there was also a sense of jeopardy, the knowledge that chaos in Russia risked their exposure. The end of the Soviet Union did not mean the end of the illegals or the desire of its intelligence services to spy on the West, though. Far from it. It was soon clear that the game went on.

A YOUNG COUPLE approached the immigration officer in Helsinki airport on April 23, 1992, and showed their British passports. The man's name was James Tristan Peatfield, from Surrey. She was Anna Marie Nemeth, from Wembley. But the immigration officer was suspicious. They seemed nervous. They had just got off a flight from Moscow but only had hand luggage and did not speak very good English. Who had won the British general election a few days earlier? They did not know. When their bag was searched, around $30,000 in cash was found inside an old shirt as well as a shortwave radio. Nemeth had some story about having been in Canada and working in an advertising agency. That was news to the real Anna Nemeth, who worked at a suburban Sainsbury's supermarket and was left bewildered when police arrived at her door. She had visited Hungary four years earlier, when her passport details must have been copied. She had never met the real Mr. Peatfield, who was from Coulsdon in Surrey. The couple at the airport next offered a cover story that they were trying to emigrate illegally and had purchased passports on the black market with money from selling women's underwear. Having such a cover story at the ready—usually involving some murky criminality—is standard practice for an emergency situation. The British intelligence officers who interrogated them in Helsinki had some hope that the woman might admit the truth, but she never did. The pair were deported to Moscow. They were illegals who had used the British identities of "live doubles" on a training mission and their failure was subject to a detailed review back in Directorate S. What it told the West was that the flow of illegals had not halted despite the end of the KGB.

The end of the Cold War had not been a moment to relax for Directorate S. Rather it was a time to double down. The world was fluid and uncertain and that meant intelligence was more important than ever. In the first few months after the coup, even while the future of the whole KGB was uncertain, a decision was taken to increase the focus on illegals. "The world is not as safe as some people present it to be," Leonid Shebarshin, head of the First Chief Directorate, explained soon after he retired. "So I think it would be a very safe measure and a good precaution to strengthen the illegal branch."

The emphasis on illegals was part of a traumatic shift by the new SVR. In the early 1990s it was forced to cut its overseas network of officers operating under diplomatic cover by a third. The fact that relations with the West were improving offered new opportunities, though. In the Cold War, it was hard for Russians to get visas to the West. Now that commercial ties were growing, spies could be placed under cover in business ventures. And the openness also offered the chance to dispatch more illegals. "While favourable circumstances exist it is essential to utilize the repose to deploy to the West as many illegals as possible and to cultivate and recruit more special agents," one member of the directorate, Alexander Kouzminov, later explained. It was almost as if, as everything else crumbled, the illegals became even more important. But they were about to be dealt a serious blow.

DRESSED IN SHABBY clothes, a Russian knocked on the door of the American embassy in one of the Baltic states in March 1992.

He was turned away disappointed, not once but twice. The Americans seem to have been nervous that the elderly man claiming he used to work for the KGB may have been a test of their new, more friendly relationship with Moscow. So the next stop for the Russian was the British embassy in Riga, Latvia, where he explained to a female diplomat that he had worked in the KGB's First Chief Directorate and had access to top secret material. The Russian pulled open his bag. Inside were spicy sausages, bread, and dirty clothes. But he also pulled out some handwritten notes. These, he explained, were the names of illegals operating in the West—both their real identities and their cover names. The British diplomat immediately called MI6. At the time, the British Secret Service was based at Century House in Lambeth, a grubby, twenty-story tower block. It was a gloomy place that reeked of the past. "Are you still here?" officials from the rest of government would sometimes joke with its chief, Colin McColl, when they saw him. The Cold War was over and people were asking what exactly spies were for. But the team that dealt with Russia, and which sat on the thirteenth floor with a panoramic view of the nearby Oval cricket ground, had no intention of stopping even though their old adversary was on the back foot. The Russian defector in Riga, they soon learned, was a prize of the highest order. Vasili Mitrokhin was a former KGB archivist who now wanted to inflict as much damage as he could on his former employers. He had secretly copied out and then buried large chunks of the KGB's operational history in the garden of his dacha.

MI6 organized Mitrokhin's exfiltration via boat on the seventy-fifth anniversary of the Bolshevik Revolution, while a young officer dug up his files so they could be removed in six large trunks. Three thousand five hundred counterintelligence reports would be sent to thirty-six countries. The US file alone consisted of eight hundred pages. The CIA counterintelligence chief rued the fact he had to travel to London cap in hand for material that could have been his. The FBI, meanwhile, said it was "the most detailed and extensive pool" of intelligence about enemy spies they had ever received. It led directly to a number of illegals, one of them a KGB man living on the American East Coast under the name Jack Barsky. He was put under surveillance by the FBI, who overheard him confessing who he really was to his wife. Disillusioned, he had actually told the KGB back in 1988 that he had been dying of AIDS so that he could give up spying and bring up his child.

Fortunately for Heathfield and Foley in Canada, they had been deployed after Mitrokhin had retired in 1984. But the defection meant that Moscow could not be sure that the legend of any illegals trained before that date had not been blown. When the scale of the disaster became clear, the SVR knew it would have to rebuild its deep cover networks by sending out even more new illegals.

In Britain, there were only limited resources to follow up the two hundred leads Mitrokhin produced about possible agents. The domestic security service MI5 had cut back its counterespionage teams by more than half between 1990 and 1994. A joint MI5-MI6 team targeting Russians in the United Kingdom had been disbanded. In one case a team was passed an intercept suggesting a message was about to be picked up from a drop site, offering the chance to catch a Russian intelligence officer in the act. But at the last moment, the surveillance team was pulled off to deal with something considered a higher priority. The sense was that all of this spy-versus-spy stuff was a bit old-fashioned. There was an important contrast in the 1990s. Western leaders and policy makers thought the Cold War had been won and Russia barely mattered anymore. Russia would "naturally" end up a liberal democratic state. That meant spying on Russia and catching Russian spies was no longer a priority. There were other threats to worry about now. But meanwhile, Russia, on the back foot, felt it needed intelligence on its old adversaries more than ever, precisely because it was vulnerable.

IN MOSCOW, CIA and MI6 officers found themselves in a strange new world. A few months after the coup, John Scarlett became MI6's first-ever head of station who was "declared" to the other side's spies. It was a second Moscow posting for the future MI6 chief and as he walked the streets, he could sense how the speed of the Soviet Union's collapse had been bewildering and traumatic. "People were out in the streets selling their personal belongings," he later recalled. Russia had gone from being a superpower seeing itself as on a par with the United States to an economic basket case. Some liberals in Moscow had hoped for an infusion of support from the West. Instead all they got were businessmen out to make money and some economists from Harvard promoting "shock therapy" in the form of relaxing price controls and privatizing state industries to create "popular capitalism." The result was a disaster. The economy went into in free fall, with prices skyrocketing, savings disappearing, manufacturing collapsing, and unemployment rising rapidly. This was a catastrophic first experience of Western capitalism in the early 1990s. For most people shock therapy simply created shock. The West saw outward signs of change in elections and privatization but it did not see the pain inflicted on ordinary people. But then it was not really that interested. The Cold War was over. It had won. And so who needed to worry about Russia?

As Western spies sat down with their old adversaries over vodka, they could sense the Russians struggling to come to terms with their new status. But in the meetings there was more drinking than trust. Each side thought the other had not stopped spying. And each was right. The SVR was certainly going through difficult times, but those who watched Russia closely in the CIA began to see warning signs that the old enemy was returning faster than anyone realized. In one officer's mind, the KGB had splintered but then reconstituted itself to fight again, a bit like the robot in the Terminator films. But no one at the top was paying attention to the warnings of people dismissed as "Cold War dinosaurs." "People looked at me like I was insane. It was lonely," recalls one CIA officer. While most people in the West believed Russia was changing, the old guard in the spy business—who could still feel the watching eyes on the streets of Moscow—never believed it had or would.

In Moscow, the Russians were likewise convinced MI6 and the CIA had never stopped their work against them. Many former KGB officers were convinced that the fall of the Soviet Union had been the result of subversion carried out by Western intelligence rather than internal decay. Now, the territory under Moscow's control had shrunk and its buffer from the West had gone. The Russians thought the strategy was continuing to try to keep Russia down when the truth was almost more painful—the West at the highest policy level had lost interest in Russia. There was no master plan. But in Moscow, the spies could also see their opposite numbers in the CIA and MI6 as busy as ever. John Scarlett's time in Moscow would end with his expulsion over the recruitment of one agent (although the real reason was a row over who his Russian counterpart would be in London). The CIA and MI6 were opening up intelligence stations in its neighborhood—including former parts of the Soviet Union like the Baltic states—which were then being used to persuade Russians to walk over the border with secrets. Russians—including former KGB officers—were lining up to sell what they knew. There were so many, the CIA was literally turning them away. The knowledge of this pained old KGB hands deeply. It was a humiliation.

In 1994, Vyacheslav Trubnikov took over as head of the SVR. Russia, he told me years later in Moscow, wanted to be an equal partner, working together against common threats like terrorism and proliferation. But instead it took the message that "Russia as the defeated side should stick to the rules and manners, which will be dictated by the victorious allies." NATO, rather than disbanding, instead expanded closer to Russia's borders, breaking what Moscow believed was a promise. Too many Western spies were engaged in what Trubnikov (rather endearingly) describes as "hanky panky." He points to one Western intelligence agency running a sting operation to smuggle nuclear material out of Russia in order to make the point that Moscow could not be trusted to look after it anymore. The belief the other side had not stopped meant you could not, either.

Inside the CIA, the tension over how to view Russia remained intense. One of his subordinates recalls Milt Bearden referring to Russia as "Ouagadougou with rockets," another remembers it being "Upper Volta with rockets," but the meaning was the same—a country that was significant only because it happened to have nuclear missiles. Bearden suggested the CIA's Moscow station would eventually be no different from Paris, staff remember him saying. The agency would gather intelligence on political developments but work jointly with the Russians on issues like terrorism, drugs, and weapons proliferation. That idea sent shudders down the spine of the small band of old hands. Deep inside Russia House, the CIA's old guard were not willing to let go. They still hungrily sought the tiniest scraps of intelligence. That was because they harbored a dark secret. The hunt for spies had not ended as the Cold War concluded. For all the talk of its adversary being on the back foot, there were a select few within the CIA who knew the Russians had moles burrowed somewhere among them.

4

"Karla"

IN JUNE 1988, the CIA's Moscow station chief, Jack Downing, had been traveling on the "Red Arrow"—the Leningrad-Moscow overnight train. In the early hours of the morning, he stepped out of his compartment to strike up a cigarette. At that moment a young Russian walked up to him, shoved an envelope into his hand, and then quickly disappeared down the corridor. This was enticing but also dangerous. The CIA man stuffed the envelope into his coat and quickly returned to his own compartment. He knew there might be cameras, so he waited until he was in a secure room at the consulate in Leningrad before he opened the envelope. Inside was a surveillance picture of himself—the type the KGB took of its targets, along with documents and a note. In the note a Russian claimed he was a KGB officer who worked in the Second Chief Directorate and whose job was to watch American spies. His marriage was on the rocks and he wanted to escape but he was willing to provide intelligence to the CIA before they got him out. It seemed a godsend after dark times for the agency in Russia.

The previous years had been bleak ones. The agency had stared into the abyss back in 1985. The CIA's Soviet and Eastern Europe division had patiently built up an impressive stable of agents over a number of years. And then they watched their assets go dark one by one in a matter of months. The morning cables brought near-daily news of disasters as KGB officers spying for America literally disappeared, often failing to show up to meet their CIA handlers at an agreed time and then missing the fallback contact. Many, they would later find out, had been executed, their families only informed that their bodies had been buried in unmarked graves.

Deep within Langley, behind a door with five cipher locks, a team had been set up to try to find out why. For some it was personal—agents whose intelligence they had handled were now dead. On a whiteboard they outlined what they knew. Perhaps it was some kind of compromise in the CIA's Moscow station? Or their communications had been broken? That was more appealing than the alternative—a traitor. In early 1986, Russia House was faced with a dilemma when a new Soviet official volunteered to spy. They took him on and had one goal: keep him alive. They kept the circle of knowledge to as small a group as possible—only a handful of people who kept off regular communications channels and who would travel roundabout routes to meet the agent. This, one veteran says, was the moment of creation of what they describe as a CIA within the CIA—an inner core determined to protect their secrets and battle the KGB. These were people who were paranoid with good reason, a kind of secret society within a secret agency. They knew they faced a formidable adversary. A foe that was determined to divert them from the truth. And this opponent had a name and, for those who met him, a face.

The Russian on the train did not give his name but the CIA code-named him "PROLOGUE." "He played the role brilliantly," says one person involved in the case. He would also, perhaps because of the way he came and went over the years, become known as "phantom." The intelligence he provided included a rundown of the activities of the CIA's Moscow station and details of the

1985 losses. He provided information that suggested that those disasters had been the result of bad tradecraft, rather than a mole. Some were skeptical about this new source but the "front office" who ran operations wanted to believe he was the real deal. The divisions over whether to believe him became bitter. There would be a show of hands on whether to trust him at one point. The decision was to keep going. There were clues that something was wrong—a piece of intelligence suggesting he was happily married, for instance, which was at odds with the story he told. But over the next three years, he would lead the CIA in a merry dance, delivering documents and diverting and distracting. He knew what the CIA wanted and he knew how to offer it to them in a way they could not resist. In 1990, the time had finally come for him to be exfiltrated. He was given false documents but never showed up for a pickup that would have taken him to a ferry to get out through the Baltics into Finland. The game was up. A cable was sent from Moscow back to headquarters with the bad news. Milt Bearden was in a room with some of the critics of the operation when he read it. "Don't you say a f-ing thing," he told them. The CIA had been played.

PROLOGUE had been a "dangle"—sent by the KGB to mislead and misdirect. His job was to keep the CIA off the scent of what had gone wrong in 1985. The KGB normally did not dangle its own officers because they knew so much. The fear was they might use the opportunity to defect for real. The Americans knew this, which was one of the reasons they took a chance on the man on the train. But the KGB was willing to take risks to protect its top spies. And the dangle was no ordinary KGB officer. The name of the Russian who had put on such a bravura performance was Alexander "Sasha" Zhomov. Two decades later, he would be present in Vienna in 2010 and watch a spy exchange he helped negotiate.

Little has been said publicly about him but there are those

inside the CIA's "Russia House" and who work on Russia at the FBI who to this day call Zhomov their "Karla"—the Russian spy who was the archenemy of George Smiley in John le Carré's Tinker Tailor novels. Other colleagues shy away from that reference. They dislike it precisely because they did not want to turn him into some elusive, mythic character. But that is what he would become. His name still elicits a pause and the drawing of breath from some veterans of the spy wars.

Zhomov was short, intellectually strong rather than physically so—one American who met him describes him as always well dressed, with smart Italian shoes. He had piercing gray eyes, thick eyebrows, and jet-black hair. Sometimes when they saw him over the years he would have an elegant mustache, other times not. He was no thug. "He is smart as shit," says one American spy who came up against him. "He got inside our heads," another says, recalling how Zhomov loved the games within games that make up espionage and the deceptions it entails. (American spies play their own games when it comes to Zhomov though, with a number trying to plant the notion he might be homosexual precisely because they know it will get him into trouble in Russia.) Those who met him often recall the silver crucifix he wore and how he would touch it occasionally, almost as if to ward off demons. In the late 1980s, he had already spent years studying the "main enemy" as a protégé of "the Professor," Rem Krassilnikov, who ran the First Department of the KGB's Second Chief Directorate. This was the department tasked with catching American spies. Zhomov supervised the surveillance teams that watched their every move. He was the man who would receive the daily reports of what the bugs in their houses had picked up or who had been out where or what they said on the phone. He knew their lives intimately. And in the PROLOGUE operation he got the chance to study his opponents face-to-face. This and other successes meant Zhomov would rise up in the KGB and then its domestic successor agency—the FSB—to eventually run the department hunting American spies and the Russians who spied for them.

One thing that marked out the Russians was their patience and persistence—not just in sending illegals abroad but also trying to catch foreign spies operating at home. CIA officers would come and go from Moscow Station; sometimes one would come back for a second posting after a decade away. But Zhomov was always there. He knew each individual, how they operated, and every trick and tactic they used. It meant he could immediately tell if there was some tiny change in behavior that suggested something was happening.

Zhomov's deception operation in the dying days of the Cold War had been designed to put the Americans off the scent of the moles that the KGB had inside American intelligence. It was not lost on anybody that the KGB would not dangle one of its own officers for nothing. If PROLOGUE was a dangle, then it meant there was likely a mole he was trying to protect. The small team of American mole hunters slowly began to narrow down the possible list of traitors who had access to the compromised intelligence. The investigation waxed and waned as superiors lost interest, but the team continued its search. By 1993, the CIA team had narrowed down the list to one officer. But there was still skepticism in many quarters and the FBI would not go along with their assessment. And so it took a spy to help push the investigation over the line.

What can seem so arcane in the world of espionage between Russia and the West is that so much of what goes on is what is known as counterintelligence—the world of spy on spy. Counterintelligence is about understanding and battling your opposing intelligence service. It can seem inward looking. But it has consequences. If your enemy has a spy in your ranks, then everything you do can be compromised. Your agents' lives are at risk and so are your secrets. Your opponent can feed in false information to deceive you. The trauma of having your intelligence service penetrated is intense. It means you have played the game and lost. It took MI6 decades to recover from the discovery that one of its most senior officers, Kim Philby, was working for the KGB (the inspiration for John le Carré's *Tinker Tailor Soldier Spy*). So how do you stop such a disaster? One way is to play defense and protect your secrets and watch your own people. But that is hard (and creates the kind of surveillance state the KGB operated at home). So the best defense is a good offense. If you recruit one of their spies you can find out who they have recruited on your side. If you feared your own intelligence service was penetrated, then the most effective way of finding the traitor is not through detective work but to recruit a spy in your opponent's intelligence service who can tell you who is responsible. It takes a spy to catch a spy. This truth drives each side to try to penetrate the other. This is the inner core of the spy world—its inhabitants sometimes drawing suspicion from their colleagues since they are looking for traitors and deception everywhere around them. It takes a particular mind-set to be able to work in counterintelligence—but for some, too much time immersed in this world of the "wilderness of mirrors" could drive you mad as you lost the ability to distinguish between deceit and truth.

IN AMERICA'S MOLE hunt, the detective work to find their mole was supplemented by vital intelligence from a KGB officer. Alexander Zaporozhsky would be one of the four men swapped in Vienna in 2010 and was one of the most important agents of the era, even though little has emerged publicly about him. Zaporozhsky had joined the KGB in 1975 and served in Africa and in the Latin America department. He was tall and athletic. Former colleagues say he was an ambitious workaholic and a risk taker. In a posting to Ethiopia in the mid-1980s, he had become fluent in the local language of Amharic and was a skilled recruiter of agents, supposedly recruiting dozens of well-placed sources. But it was not the most prestigious posting and, unhappy over his career and money, he offered to spy for the CIA. He was code-named "Max." The Americans found him well-spoken and self-assured. "If you were casting someone for the role of a KGB colonel, he would get the part," says one person who knew him. But he had been hard to handle, sometimes feeding erratic information to his handlers and disappearing for long periods. Eventually he would become a colonel in the SVR's North American department. In the years of turmoil and decline in the early 1990s his disillusion grew. These were difficult times with the country in crisis and spies were eager to cash in as discipline deteriorated. In 1993, Zaporozhsky, even though he did not know the traitor's name, passed on a piece of information that helped identify the mole that the KGB and SVR had inside the US intelligence community. For this clue, Zaporozhsky would be well rewarded but also pay a price.

The clue confirmed that Russia House itself had a traitor. Aldrich Ames was a second-rate spy who was convinced he was first-rate. For all his failings, he had still been appointed to a sensitive role working on counterintelligence in the Soviet division. In 1985, he had used the pretext of trying to recruit a Soviet diplomat to instead offer the KGB his services. In his second meeting he provided a haul of information the KGB would describe as stunning, tilting the balance of the spy wars in Moscow's direction. "He was as lazy, sloppy, and careless at spying as he was at his CIA job," writes former colleague Mike Sulick. When investigators closed in, they found operational notes torn up in his trash. On February 21, 1994, the FBI arrested him. The revelation that the CIA had harbored a mole working on its crown-jewel operations was a hammer blow.

"His treachery devastated the CIA," Sulick later wrote. A generation of Russian agents had been sold out and now it looked like the agency could not be trusted to keep secrets. Ames's treachery also played into a debate in Washington that asked what the CIA was really for now that the Cold War was over. Congress and the press went on the attack. Worse, a few on the inside soon realized Ames could not have been responsible for all of the compromises. That meant there was another traitor still out there—this was the man who had made the drop in the park as the Russian coup was beginning on the Monday morning in August 1991.

A CIA delegation headed to Yasenevo to deal with the fallout of Ames. There were no pleasantries as both sides tried to work out how to deal with the other in this new world of liaison meetings but when both also knew the old spy games continued. "These things happen," the head of the SVR said, as if to brush events off. "It's the nature of the business." The CIA man explained there would have to be consequences—there was too much anger in Washington. "Things will get ugly. That's inevitable. The question is how can we contain the damage?" Each side expelled spies from the other's embassy. It was agreed they should try to maintain contact, though. Soon after the Americans were introduced to the man who would be their new liaison officer. The door swung open. In walked Sasha Zhomov.

THE CAPTURE OF Ames, in turn, started a spy hunt in Moscow. The FSB was determined to find out why their star agent had been discovered. Battered by defections among their ranks as their country went through a crisis in the 1990s, Russia's spy hunters felt a burning anger. They had held the upper hand but had now been betrayed—including by their own colleagues. And they wanted revenge. Zhomov would be the man to deliver it. It would become his driving mission for the years to come. He never believed the false trail the CIA had put out that Ames had been caught purely thanks to detective work. He knew that spies catch spies. When one of your agents is caught, it means you have been penetrated. And to find that spy, the best method is, of course, to recruit a spy in your opponent's camp. And so the wheel turned again and the cycle of espionage began over. The ghosts of the Cold War would continue to haunt the corridors of the Lubyanka and Langley even as that conflict faded, the mole hunters on each side determined to exorcise the demons of the past. But in doing so they would raise new ones for the future.

5

Undercover

"It is like walking a tightrope," Donald Heathfield said of his double life as an illegal. "When you first step on it, you are scared. It is very high, you don't know what to do. But if you keep doing it in the circus for ten years, you can do it with your eyes closed." It took a full decade to learn the ropes, a decade of patiently building your cover, the thrill of clandestine work mixed with the mundane reality of everyday life. It was not glamorous. When Heathfield set up his own business in Canada, it was not the "Universal Export" of a James Bond front company. Instead, the deep-cover Russian spy drove a van full of baby diapers door-to-door. "Diapers Direct" sold a box of two hundred for twenty dollars for those who valued the price and convenience of a bulk order delivered direct to your house. Heathfield had a good feel for the capitalist West and the business generated publicity with a feature in the *Toronto Star* newspaper. At the same time, he was also taking a bachelor's degree in international economics from York University. His wife, Elena, now Ann Foley, joined a Catholic church, its customs and rites having been taught in spy school. One thing they needed was people who could vouch for them as friends and perhaps sign the right documents. Foley went to a church where the priest was an elderly, warmhearted man. Foley first joined the choir and used that to get herself close to another young woman at the church who in turn introduced her to the priest. Every relationship served a purpose. Their family was also growing. On June 3, 1994, a second son, Alex, was born in Toronto.

For Directorate S, children posed a dilemma. In the Cold War, if one half of a couple remained in Russia there were special boarding schools for the children of illegals. In one case an illegal spent seventeen years abroad. His wife remained in Russia and his son would be brought to meet him in a Western European country the illegal could visit. This was "so that the boy saw what a worthy father he had," Yuri Drozdov, a head of Directorate S, later said. "But a tragedy happened. The son, on vacation at camp, drowned, and the father came to the funeral for a day. One day." That was all the time that was possible while retaining his cover before he had to return to the West.

Heathfield and Foley hid their old selves from those around them—including their children. They spoke little of their early life. Every family has its own quirks, and no one has grown up in another to know what is entirely normal and what is odd. At one point the children said they remember meeting people they thought were grandparents on vacation somewhere in Europe. When asked if they were speaking Russian, Alex replied, "I was really young, I have no idea." Timothy says he remembered seeing them every few years until he was eleven. "If I had seen them when I was older, I would have realized that they don't speak English—they don't seem very Canadian." The children would get Christmas presents from these grandparents. They were told they lived in Alberta, Canada. Photos showed them with snow in the background. It was really Siberia.

The life of an illegal involves sacrifice. Contacts with family back home were limited. The couple's parents only learned of the arrival of their grandchildren weeks after the births. The parents had to send letters to a mailbox in Moscow, where they could be encoded. Sometimes when Heathfield and Foley were abroad and could meet KGB and then SVR officers in person, they would get an original handwritten letter. Once it was read, it had to be destroyed. Every three years or so they might be able to meet their family on a trip but there could be no promises about when it might happen again. Bezrukov's father would die while he was undercover. He only learned about it weeks later. There was no way he could go to the funeral. "You must have a strong a core," his wife would say. She missed the chance to pay last respects to her sister before she died. "It is a heavy burden for an average person. You have to be very strong and very certain you are doing the right thing."

The parents were also prepared for questions from curious Americans. In the Soviet Union, citizens were given vaccine injections in a different place than Americans and it left a scar. Elena—now Ann—would try to keep the scar covered. She would have to explain she had grown up in an African or Asian country where it was done like that. In the Soviet Union, dentists also used a kind of cement filling for teeth that would never be used in America or Europe. The same excuse—early years spent in the developing world—would cover for that. Every detail mattered. When their predecessor, Konon Molody, had been caught in the United Kingdom, one thing that gave him away was the fact that he had not been circumcised. A doctor recalled that the real Gordon Lonsdale, born in Canada, had been.

For illegals, a double life meant long hours. There were full-time jobs to hold down plus the normal struggles of family life. And then there was the second career as a spy, seeking out contacts and composing messages to Moscow Center. Every time an operational act had to be carried out—a letter marked with invisible ink that had to be posted, a signal site checked—an illegal would have to spend at least two or three hours on foot and on public transport carrying out a surveillance detection route to make sure they were not being tailed. The couple was disciplined about never speaking Russian in the house. "Heathfield does not know Russian," Bezrukov would later say of his old self. "If you wake him up in the middle of the night, he doesn't speak it." Nor would they discuss anything operational—just in case there was a bug. The same went for talking in their car. If they did need to talk, they would go for a walk in the park, often taking the kids with them; later they would go jogging together. Their old selves were buried deep. To adapt to the life in a foreign country, you have to give up everything you had from your childhood, "forget it, get rid of it," Foley later said. "Otherwise, you will suffer from nostalgia. Any mention of Russia, Russian music, Russian speech in the street, throws you off the balance and provokes memories." They created false memories drawing on real people or events and transferring them onto new people and Canada. At some level, they began to believe these were true and that they had indeed lived a different life. But there was always the knowledge in the back of your mind that one day something might happen that would knock you off your tightrope and send you tumbling down.

In August 1995, the family sold off the business and house and moved to Paris, where Heathfield studied for a master's degree in international business, living in a small flat near the Eiffel Tower. The same year that they left Canada, Alexander Poteyev, now an SVR officer on the rise, moved to New York. For now, he was a loyal member of the Russian intelligence service despite all the problems his country was enduring.

From afar, Heathfield and Foley could read the news about the turbulence back home in Russia in the 1990s. In their absence, their homeland was changing rapidly. These were difficult years as the country struggled to come to terms with its new status and economic crisis. Everyone was on the make, including the spies. Organized crime began to emerge, settling its scores with violence. A few sharp-eyed Russians bought up shares in privatized industries and accumulated huge wealth. They would become known as oligarchs. In the wild-west capitalism of the 1990s, wealth and power were tightly bound and a gun to the head settled disputes.

In the early evening of June 7, 1994, a remote-controlled bomb detonated as Boris Berezovsky's Mercedes drove away from his office. The driver was killed but the target survived. A mathematician by training, Berezovsky had started a car dealership but then moved into everything from oil to TV to airlines. A brooding and pugnacious character, he would become the first among equals of the oligarchs and a power in the land, always ready to scheme and plot. A sandy-haired, serious-minded, thirty-one-year-old FSB officer named Alexander Litvinenko was assigned to investigate the assassination attempt. Litvinenko had been recruited into the domestic arm of the KGB in the late 1980s. In 1991, he had been sent to Moscow to work on organized crime. That gave him an education in how the security services had intertwined themselves in the new chaotic, corrupt, freewheeling economy. As well as investigating the Berezovsky assassination attempt, Litvinenko was also ordered to report back on the oligarch to the FSB. Later that year, a brutal conflict broke out in Chechnya and Litvinenko was sent to work on counterterrorism. His and Berezovsky's paths would cross again a few years later. At one point the chaos was so bad that Boris Yeltsin looked as if he might lose the 1996 election—to, of all things, a communist. Yeltsin arranged a secret deal with the oligarchs. They would throw their money and influence behind his campaign in return for stakes in the vast state-owned natural resource industries and also more influence. Berezovsky organized the deal and became deputy head of the national security council in the wake of the election. As the decade came to an end, his power was reaching its zenith. But it would not last much longer.

ON AUGUST 20, 1999, more than a decade after first surfacing in Canada, Donald Heathfield and Ann Foley finally arrived in their "target" country. Heathfield and Foley had come to America—and would become naturalized American citizens—for a reason. Human intelligence is about people—people who have access to secrets, to power, and to influence. The job of the illegals was to find those people. "Our goal was not to steal a blueprint—as they show it in the films," explained Elena Vavilova. "Our goal is to find that 'somebody.' " That could be a diplomat or an engineer, a politician or an academic. The illegals' mission was to subvert America from within, infiltrating deep into its society and in doing so identifying and helping recruit people who could aid Moscow.

Heathfield had landed at one of the best places for his particular line of business. Like any student, when he applied to Harvard, he knew it was a ticket to the big time, a chance to make contacts and open doors, to work his way into the elite circles of American life. The difference from most other students was that this would be for Moscow's purposes and not just his own self-advancement. He was studying for a master's in public administration at the Kennedy School of Government. The school was often looked down on by the traditional academic departments, but everyone knew the reason to attend was the connections it offered. The faculty was packed full of former officials and politicians. Former and current CIA officers, senators, and policy wonks all made it their home. And the students who came were ambitious and would make their way into government in the United States or around the world, especially in the midcareer program that Heathfield joined.

Heathfield arrived in late 1999 and gained a reputation as a sociable member of his class of two hundred students. He said he was from Montreal. If pressed about his accent, he would say he was the son of a diplomat and had been to school in the Czech Republic. Heathfield organized a drinking night in which fifty members of the class visited Canadian students to try high-end Scotch. "We called it the Royal Canadian Scotch Stagger," one later remembered. The night ended at 3 a.m. at Heathfield's house. The networking opportunities were ample. Among those in his year was Felipe Calderon, who would later become president of Mexico, as well as others who would run for political office in the United States and work with the US Army. Heathfield, fellow students remember, was particularly good at keeping track at what people got up to afterward. Harvard would provide the credentials for his future career but was also the first stage in trying to find that "somebody" who could—wittingly or unwittingly—serve Moscow. "The main task of an agent is constantly climb the social ladder, achieving contacts with more and more prominent society members. Because it is only there that you find really valuable information," he later said.

On a sunny day in May 2000, Donald Heathfield attended his graduation ceremony. He was all smiles. It was a proud day for all the family. Ann was at his side, looking smart with a pale blue jacket and a pair of sunglasses. At her side were young Alex and Timothy. The mission seemed to be progressing well. But what the family did not know was that they were being watched. Spotting an illegal in the wild is incredibly challenging—that is the point. They could be anyone—your neighbor, your coworker (even, in one case, your dentist)—they are almost impossible to find. Unless, of course, you have your own spy in their ranks.

Heathfield and Foley were ghosts—their identities stolen from the dead. But they were not invisible. Close to them at the graduation ceremony—just a few feet away—were people silently hovering around them. These figures furtively snapped pictures of the couple with their children. They were members of the FBI's Special Surveillance Group—the SSG, often called the Gs. Decades earlier the FBI's surveillance had been something of a joke—the regulation dark suits and white shirts were a giveaway. So they had learned from their British cousins at MI5 how to set up specialist teams who could blend in anywhere and looked like ordinary people—a mirror to what the illegals themselves were trying to do. And because their job was to be unseen as they tailed their targets, the Gs were also known as "Ghosts." That sunny day at Harvard, ghosts were chasing ghosts.

6

The Source

In the late 1990s, as Donald Heathfield was arriving at Harvard, Alexander Poteyev was making his way every morning to the Russian mission to the United Nations. The mission is a slice of Moscow dropped slap-bang in the middle of Manhattan, a dull-gray Soviet-style twelve-story building at 136 East 67th Street. On the same block sits the 19th precinct station of the New York City Police Department and the 16th Ladder of the Fire Department. Directly across the road is a synagogue. In an act of defiance during the Cold War, the synagogue had a large plaque placed outside for Russian diplomats. "Hear the Cry of the Oppressed. The Jewish Community of the Soviet Union," it reads. No doubt, the FBI also has a presence somewhere to watch those leaving and entering the Russian building. And in the late 1990s, Poteyev was of particular interest. He is an elusive figure whose life is deliberately shrouded in mystery. But while FBI and CIA officials adamantly refuse to confirm or comment on the identity of the origins of the investigation into Russian illegals and many of the details come from Russian sources, there is no doubt that Alexander Poteyev was the key figure. He was the reason why Donald Heathfield was followed at his graduation. It took a spy to catch the spies. Poteyev was "the source."

. . .

POTEYEV WAS BORN on March 7, 1952, in the Brest region of Belarus, in what was then the Soviet Union. His father, Nikolai Poteyev, had commanded T-34 tanks with distinction in World War II, earning the title "Hero of the Soviet Union" for his role in battle in the Baltic front in September 1944, but died when Alexander was twenty. His son followed him into the army and in 1975 he joined the KGB, first serving in Minsk and then from 1978 in Moscow, though not yet as a spy. During these years he met and married Marina and they would have a daughter and then a son.

In 1979, Poteyev went on the Advanced Course for Officers (KUOS). The name was misleading. It was better known as "the school for saboteurs." This was the KGB's elite paramilitary training for those destined for either irregular combat or stay-behind and sabotage roles in the event of World War III and other conflicts. It came under the wing of Directorate S's Department 8, which carried out "special operations." These were unique roles that combined the ability to work undercover for long periods as a spy with high-end military training. Poteyev would soon get his chance to put those skills into action.

On Christmas Eve 1979, Soviet troops were ordered into Afghanistan to quell a growing insurgency and topple a leader insufficiently pliant to Moscow's orders. Directorate S's Department 8 had already tried to assassinate the Afghan leader by sending an illegal documented as an Afghan to get a job as a cook and poison his fruit juice. But the Afghan leader was suspicious enough to carefully mix his drinks. So Moscow resorted to an all-out assault on the presidential palace led by Directorate S and GRU special forces (the "illegal" cook had to hide to avoid being shot by his compatriots during the raid). Poteyev was sent early to Afghanistan as one of the Directorate S "Zenit" teams. The invasion set the scene for a brutal struggle as a bloody insurgency by mujahideen fighters gathered strength. This was fueled by weapons and money sent by Islamic states and the CIA (led, in part, by Milton Bearden), who saw it as a chance to give their Soviet enemy a bloody nose. It was a savage and dirty fight—the Soviet Union's version of Vietnam—from which many young men returned in body bags. Poteyev was on the front lines. One of the only photos of him is as part of Zenit—a young-looking figure in fatigues with a blank expression. One KGB colleague from those days simply remembers Poteyev as someone who liked to drink and who had a good sense of humor.

Poteyev next served in "Cascade," a new task force formed to fight guerrilla warfare. Like a US special forces team in Afghanistan in later years, their job was to seek intelligence and use it to hunt for the enemy and its agents in the towns and countryside. This included using illegals who could pose as fake mujahideen to lure others into ambushes. Out of Cascade, Yuri Drozdov, the head of Directorate S, would form a new special forces unit for covert action behind enemy lines called Vympel. The intervention in Afghanistan proved disastrous for the Soviet Union, but for Poteyev, it was a stepping-stone. He was awarded the Red Star and other decorations before being selected to attend the Red Banner Institute to be trained as an intelligence officer (at the same time as Bezrukov and Vavilova were being trained as illegals). Poteyev must have been effective as a spy, because in 1995 he was sent on a plum posting to New York. He arrived as a second secretary at the Russian mission to the UN and stayed through 1999. The Russian ambassador to the UN at the time was Sergey Lavrov—later Russia's long-serving foreign minister.

When he arrived every morning at the mission on 67th Street, Poteyev would take an elevator up to the eighth floor. In the lobby were two steel doors. There were no signs, but one was for the *rezidentura*—the intelligence station—of the GRU, military intelligence. The other door led to the SVR's *rezidentura*, where Poteyev worked. Officers would pull out a small piece of metal and touch the head of a screw in the lower right corner of a brass plate by the door. That would complete an electric circuit—sometimes giving the person a tiny jolt—and the bolt would slide open. Behind the door was a cloakroom. Under the watchful eye of a camera, everyone would have to leave their coats as well as any electronic items in a locker to make sure nothing could be smuggled in or out. Next came another solid steel door with a numeric lock that required a code. Beyond it, the GRU and SVR offices were known as "submarines" since they were tightly enclosed to prevent FBI surveillance. Special teams had been flown from Moscow to create the structure, which sat on top of springs, creating a space from the main building structure to prevent cameras and listening devices being inserted. The walls were several inches thick and coated with wires that vibrated to emit a white noise. There were dedicated electrical and ventilation systems. On the SVR side there was a corridor ninety feet long with offices on either side. These were given over to the different "Lines" with different roles—one carried out technical interception; another, Line X, looked for technological secrets; VKR studied American intelligence; PR sought political contacts and information. Poteyev would walk into an office in the far corner of the floor.

Poteyev was a Line N officer. These are the spies under diplomatic cover whose job is to support the work of Directorate S illegals. Sometimes this can mean doing the legwork to establish a false identity—they were the ones traipsing around graveyards and church registries looking for names of dead children. They would also be in charge of getting hold of documents or visa applications so that Moscow Center can produce convincing forgeries. It could also mean supporting illegals directly. The whole point of illegals is that they do not appear Russian, so direct contact with anyone from the embassy is kept to an absolute minimum.

But there might be moments when some kind of indirect contact is required—it could be an emergency signal left somewhere if something is going wrong—that might require a Line N officer checking every week to make sure there is not a chalk mark at a particular place. Or there might be documents or cash to leave in a dead drop for an illegal to pick up. Line N officers might also pass off new documents to an illegal transiting through their country or they might meet an illegal in a third country (often Mexico or somewhere in Latin America for US-based illegals).

There were 60 SVR officers based in New York in the late 1990s. They were running about 150 sources. As well as the mission, there was the Russian Consulate at East 91st Street, close to Central Park. Poteyev would likely have lived in a large, dingy tower on West 255th Street, in Riverdale in the Bronx, that was the residential compound. The rooms were small and smelly, with plenty of cockroaches. There was a small bar with cheap liquor and cigarettes, even a sauna, a swimming pool, and a school so that the Russians were not too tempted by the bright lights and enticements of the city on their doorstep. Surrounded by a chain security fence, the compound had been built on a steep hill. This allowed antennas to be placed that could intercept communications, and on the nineteenth floor, Line VKR—foreign counterintelligence—ran a system called "Post Impulse," which tracked FBI signals. If they saw several FBI signals in the vicinity of one of their SVR officers they knew they might be under surveillance. In New York, the top mission for the SVR spies was penetrating the US mission to the UN; next was the missions of other permanent members of the Security Council—the United Kingdom, France, and China, followed by Germany and Japan and other NATO countries. Next were New York financial institutions, then the universities like New York University and Columbia, and finally Russian immigrant groups and foreign journalists.

From the day he arrived at the Russian mission to the UN, Poteyev had been closely observed by the FBI. They did this to all Russian officials using observation posts, surveillance teams, and bugs. But the FBI and CIA had a particularly good insight into what was going on at the Russian mission to the UN while Poteyev was there. That was because they already had a spy on the inside. Sergei Tretyakov was technically a first secretary but actually the deputy SVR resident since his arrival in 1995. A rotund, outgoing character, Tretyakov was one of those who looked at the SVR in the mid-1990s and saw only decay. When he visited Yasenevo, he remembered how it had grown grimy in his absence. The bathrooms, once spotless, now looked more like you would expect in a railway station. People looked unkempt and were drinking and leaving the office early. Jobs had been cut and many had gone off to make money. Disillusioned, he began working for the Americans and would spend the late 1990s providing highly valuable intelligence, including the names of undercover SVR officers across the United States and their agents.

The two Russians would have known each other, but Tretyakov did not introduce Poteyev to the Americans. Keeping agents separate was a vital principle of tradecraft or else the risk would be that one of them—if discovered or turned—could compromise the other. Tretyakov would, though, have been able to tell the Americans all about the officers in the mission, including Poteyev, which may have aided their understanding of him and whether he could be approached. In October 2000, rather than return to Russia, Tretyakov simply disappeared. SVR operations in North America were dealt a huge blow. But the SVR did not know things were even worse than they feared. There was another spy.

Poteyev was watched for some time by the FBI. FBI teams study every Russian diplomat, building up a file on them. What is their work pattern? Does their routine make them look like a real diplomat or might they be a spy? What kinds of things do they do in their spare time? As well as hoping to catch them in the act of espionage (always difficult), the counterspies are looking for "the hook"—the aspect of their life you can cast your line toward and hope it catches so you can reel them in. Sometimes they will be overheard on the phone talking in a way that sounds like high-minded ideological disillusionment but other times it is because they are observed gazing longingly at large plasma screen TVs in shop windows as they walk downtown.

What motivates such people to turn against their country and spy? Occasionally in the Cold War there were genuine ideological turncoats. But money and general disillusionment were more of an issue for Russian spies in the 1990s. They had watched the ideology they had signed up for disappear and their savings evaporate. Some literally became chicken farmers. Meanwhile, they could see others back home cash in in the new world of crony capitalism. Why—after all their service—should they not have some little nest egg to prevent their family from struggling? That was one reason. But the truth is that simple answers rarely suffice. Each case is unique. If you speak to those who target Russians, they say the simple notions of motivation rarely apply. The reality is much harder to unpick. It is sometimes tempting to reduce it all to something like money or grievance or ego. But, one old hand explains, the Russians are complicated. They are all maneuvering in their bureaucracy against each other, sometimes sleeping with each other or their partners, collecting compromising information on each other, and holding grudges for some slight inflicted on them years ago—any of which could lead one of them to suddenly decide to turn. Another former spy has a different take. In post-Soviet Russia it was not the people you expected to turn who did. Rather than unhappy, low-level intelligence officers, it was more senior ones who changed sides. They had got far up the tree but then realized they were not going to go any further since when you reached a certain level, politics and corruption took over and it did not matter how good you were. That was the moment you might be willing to turn.

What was the case with Poteyev? Russian spies would later bitterly attribute Poteyev's actions to the "unraveling" of the 1990s, when everything was for sale and when security was so lax and no one cared where you got your money or stashed it away. The 1990s were difficult times for Russia's spies—the old certainties of communism gone, a new, almost alien world back in Russia in which a wild form of capitalism and gangsterism seemed to be flourishing. For old SVR hands, a demoralized service without an ideological compass was vulnerable to its opponents, allowing MI6 and the CIA to have a field day. His former colleagues would claim Poteyev sold them out "banally" for greed, saying the Americans exploited his love of money and alcohol. They would say that he had got fond of life in the United States and its luxuries. There would be talk of shady deals in which Poteyev was involved in money laundering and helping other SVR officers buy homes and move their cash to America. All of this was used by the CIA, the Russians would later say. Much of this was wrong, misinformation, or reflected the bitterness of betrayed colleagues. Here, for the first time, are the outlines of the story.

The recruitment of Poteyev took place in 1999, at the end of his posting. It was not by the CIA but by the FBI's New York field office. Recruiting Russian intelligence officers inside the United States is the province of the FBI rather than the CIA. In general, the FBI's job is to catch people breaking the law and the CIA's job is to break the laws of other countries by stealing their secrets. One side is cops, the other robbers. They have different cultures and relations can be rocky. There was real tension in the late 1990s between the CIA and FBI over counterterrorism, but the relationship in New York on counterintelligence was tighter. New York—with the world of business as well as the UN—was a fertile hunting ground for the bureau's officers seeking to recruit Russian assets and the FBI's New York office was big enough to have a critical mass of counterintelligence expertise. "The New York field office is its own world—with its own worldview," one former FBI counterintelligence officer explains. "Most New York agents believe the sun rises and sets in the New York office," says another. With so many potential targets in the city, the office had experience and swagger.

The field office is housed at Federal Plaza, a few blocks from the site of the twin towers of the World Trade Center. The counterintelligence team in the late 1990s was housed on the twenty-sixth floor, part of the National Security division, which was headed by the larger-than-life John O'Neill. In 2000, he mislaid a briefcase for a few hours that contained details of counterterrorist and counterespionage cases (perhaps including that of Poteyev). The briefcase incident provided one more excuse for those who did not like his hard-charging style, and he left soon after for a job at the World Trade Center, where he died on September 11, 2001.

The FBI would have seen that Poteyev liked life in America. But that was not a reason to assume he'd be open to spying for the United States. He was pitched by the FBI to see if there was a chance he might turn. This was commonplace. And it had happened with him not just once but again and again. He had declined. But as his posting was coming to an end and he was about to return to Russia, something changed. He decided he was ready. Why? It seems to have been a mix. There was certainly money. Like others in Russia in the 1990s, his pension had been cut drastically. There was also some disillusionment at the way the SVR had acted back home, including on a personal level, failing to support him through some difficult family times, including the death of a relative. But there was also disgruntlement. He wanted to extend his tour in the United States, but his request had been denied by Moscow Center. Poteyev was not quite what is called a "walk-in"—someone who walks in off the street and offers him or herself out of the blue. Rather, it was as his time in New York came to an end that he changed his mind and indicated he was interested. The Russians believe he was recruited in June 1999. US officials will not comment.

It was precisely what had annoyed him—having to go back home—that offered a rich opportunity for the FBI. Tretyakov had defected and stayed in the United States, but the real prize was being able to run an agent-in-place in Moscow who could continue to work his way up the system and deliver secrets. That is what the FBI wanted and the risk Poteyev was willing to take.

To succeed, an intelligence organization makes sure the recruitment and running of an agent is on a need-to-know basis within its own corridors. There is a reason for that. And the Poteyev case was a prime example. Just as the FBI had, in the form of Poteyev, recruited a source inside the SVR, the SVR was at that time still running its own agent inside the FBI who had not been identified. Robert Hanssen was the man who had been dropping off secrets in the park on August 19, 1991, as the Moscow coup took place. A misfit who shared details of his sexual fantasies about his wife online, he began to spy way back in 1979. He had actually stopped in the wake of the coup and the end of the Soviet Union, but in 1999, he resumed contact with the SVR and began to provide more vital intelligence. If word had got around the bureau about the new recruitment, Poteyev may not have lasted long. The counterintelligence spy games of the Cold War had still not yet finished playing out.

As he landed back in Moscow, Poteyev would have known his new secret could destroy his life if it was discovered. He was heading for a double life and walking his own tightrope, one that he knew could end at any moment if he made a slip or—more likely—if someone else betrayed his secret to the SVR. But on his return, he received a dream posting—for him and his new American friends. He was to join the senior ranks of Directorate S. He would eventually become the deputy head of Department 4—the team responsible for running illegals in the United States, Canada, and Latin America. This was one of the most secretive, compartmentalized parts of the entire SVR. Only a tiny group of people was allowed to know the identities of illegals sent abroad, in order to protect them. Only three officers had access to the personal files of illegals operating in the United States. Poteyev was going to be one of them and he was in charge of their day-to-day operational management. This was a stunning success for his handlers—if they could keep him from getting caught.

RUSSIA'S SPIES WERE on the back foot. And at the same time that the United States was scoring a success with one veteran of the Afghan war, MI6 had managed something similar with another. And the fate of the two spies would ultimately be drawn together.

IN MADRID IN the summer of 1996, a rugged first secretary at the Russian Embassy was walking in a park with a businessman from Gibraltar. The first secretary was an officer of the GRU—military intelligence. His name was Sergei Skripal. The businessman was an MI6 officer operating under "natural cover." The Russian had been spotted by the Spanish as someone who might be interested in money. He was first introduced to a Spaniard. They talked about going into business together, exporting wine to Russia. The Spaniard next introduced Skripal to the businessman, who would offer something more lucrative but also more dangerous. The lure was the promise that together they might be able to go into the oil business in Russia. But as Skripal prepared to return to Moscow at the end of his posting, the MI6 officer showed his hand and revealed what he was really after.

This was typical of the kind of pitches to Russians in this period. It would involve approaching someone and offering them a contract for consultancy or discussions about business. No secrets would pass. But after a while, it would be explained that this work was sadly at an end but there was another possibility—perhaps further business dealings of a more sensitive nature. This would require the individual being put in touch with someone more closely associated with the British government. If you pitch right away saying, "Do you want to pass secrets for money?" the answer will be no. But one step at a time, reeling someone in can be a lot easier.

By the middle of the decade, MI6 had reduced by two-thirds the amount of effort it put into spying against Russia and the former Soviet Union. But a sharp young officer, Charles Farr, had taken over Russian operations in London and made the case that MI6 needed to take advantage of the moment and recruit sources for the long term. The officer who met with Skripal was a prodigious pitcher of Russians; "a magic recruiter" is how one of his colleagues remembers him. Skripal proved receptive. The reason was simple—he liked money and did not have much.

Skripal grew up in Kaliningrad, a strategic enclave on the Baltic coast sandwiched between Poland and Lithuania. It was a closed military zone but near enough to the West for the young Sergei to pick up the tinny sounds of BBC World Service radio, which carried news of a more colorful world. His father had been an artillery officer and Skripal joined the elite Soviet airborne troop. He, like Poteyev, had been sent to Afghanistan at the opening of the conflict to carry out undercover missions. These included targeting locals suspected of working with the CIA. Where Poteyev worked for the KGB, Skripal was selected to join Russian military intelligence—the GRU. The eyes and ears of the General Staff, it had always been a tough and uncompromising service, motivated more by patriotism and a military ethos than the ideological focus of the KGB, with which it competed.

The GRU took treachery seriously. There were claims that recruits were shown a film of one traitor being pushed into a furnace and burned alive. It was the one part of the Soviet Intelligence apparatus that did not change in name or culture as the Soviet Union became Russia. It included special forces and its own illegals whose job was to prepare for sabotage in the event of war. Caches of weapons were left ready (the West did something similar in Western European nations in case they were overrun). The GRU also had officers stationed under diplomatic cover in embassies, collecting information on military intentions and technology, including by recruiting agents. Skripal's role, after graduating from the Diplomatic Military Academy, was in its First Directorate, which focused on Europe. After the collapse of the Soviet Union, he had ended up in Madrid. And by the end of his time there he had acquired an MI6 code name: Forthwith.

How important was he as an agent? Opinions differ. In the late 1990s, the customers of MI6—the Foreign Office and Ministry of Defence—were primarily interested in political developments in Russia and details of proliferation of weapons of mass destruction. But what Skripal could offer was largely counterintelligence—details of the GRU and its operatives. This was of niche value, although for those interested in it, Skripal was an excellent source. His cooperation was erratic rather than regular. He was not classed as a full-fledged, recruited agent when he left Madrid and no one was sure his cooperation would continue when he returned to Moscow. But it would. And his new, senior role in the personnel department meant he was able to identify hundreds of

GRU officers operating under diplomatic cover overseas, whose details MI6 could then pass on to other countries.

By the start of 2000, Skripal and Poteyev were both in Moscow, providing intelligence from inside their own spy agencies. The actual recruiting of a Russian spy on their home territory of Moscow itself is almost impossible. The slow cultivation of a relationship and the careful conversations required to sound someone out would almost certainly be spotted by the vast counterintelligence machinery. But if you have managed it overseas—like with Poteyev in New York or Skripal in Madrid—then it may be possible to run them back in Russia. But only one of this pair would escape capture. With Poteyev in Directorate S, US intelligence had scored a stunning coup. They had opened up a window right into the heart of the most secretive part of their adversary's operations against them. As long as they had their source in place, they would be able to track illegals operating in the United States.

BUT THE RUSSIA that Poteyev had returned to after his time in New York was about to change. After the chaos of the 1990s, a new power was rising in the form of a former KGB officer who had traitors in his sights. The United States had its window into the illegals program. But how long would it last?

7

The Investigation

Rifling through someone else's safety deposit box is the province of two kinds of people—thieves and FBI agents. On January 23, 2001, the latter were at work in Cambridge, Massachusetts, the leafy home of Harvard University. They were covertly searching through the personal items of one of the university's recent graduates. They were there because, thanks to their source in Moscow, they believed the owner was a Russian illegal. Inside they found a birth certificate. It was for a Donald Howard Graham Heathfield. They snapped a photo and quickly returned it. It would take another four years for another piece of the puzzle to fit alongside the birth certificate. "Suddenly but peacefully," Howard William Heathfield died at his home in Burlington, Ontario, Canada, on Thursday, June 23, aged seventy, a 2005 death notice in the Canadian press read. "Howie" left behind a wife, three children, two grandchildren, and two dogs, called MacGyver and Holly. There had also been a son, Donald. But Donald had predeceased Howard. Although the middle name was different, both the death announcement and the birth certificate listed the mother's name as Shirley. Donald Heathfield was not who he said he was.

Investigating illegals—like being one—required patience and attention to detail. Inside the safety deposit box, there were also photographs of Heathfield's wife, Ann Foley, when she was younger. Many people have their memory boxes—the keepsakes, photos, and letters to recollect an earlier life. But for an illegal to do so was dangerous since it was, almost literally, another life. It was a security lapse. Perhaps Foley needed something to cling to in order to remind her who she really was. Not Ann Foley the Canadian, but Elena Vavilova, the Russian from Tomsk. But she had made a mistake. Surviving as an illegal is all about details—tiny details. And Foley and Moscow Center had missed one. Stamped on the negatives was the name of the company that had produced the film. It was called TACMA—a Soviet film company. It was another piece of crucial evidence for the FBI as the investigation got under way.

It is a mistake to describe the illegals arrested in 2010 as a "spy ring" or network. That implies they were one group working together. The reality was that they were sent out in pairs or individually at different times—some deep in the Cold War, others toward its end, and some after it was over. They would have been aware there were other illegals in the country, but for reasons of security they would not know who they all were. The SVR would not want the discovery of one illegal to allow the FBI to find the others by following them. But for the FBI, this was not a case of finding one illegal and then following them to another. Thanks to Poteyev, their source in Moscow, they knew who was in the United States and who was coming.

The first act that the FBI monitored had come a year before the safety deposit box. On January 14, 2000, Vicky Pelaez made a trip from New York to her native Peru. There she met a Russian official in a public park. She was given a bag. Inside was money from the SVR. What she did not know was that two FBI agents were videotaping the whole show. Once that was done, she called her house in Yonkers, New York. The FBI were also listening in on the line. "All went well," she told Juan Lazaro. Vicky Pelaez was neither Russian nor an illegal but she was married to someone who was both. And she has always maintained she did not know her husband was a KGB illegal whose career had begun deep in the Cold War.

PELAEZ WAS A dark-haired, charismatic Peruvian. She was not a trained spy and her marriage was not arranged by Moscow Center. And rather than keep a low profile, she had lived a life marked by drama, controversy, and an outspokenness that did little to hide her political views. Born in 1956, Pelaez had studied journalism and became one of the first female reporters in Peru, working first for newspapers and then TV. Stylish and brave, she broke down barriers and challenged the stuffy style of traditional news. She quickly built a reputation as a gritty journalist, unafraid to take risks, and knew how to insert herself into a story, slipping behind police cordons and getting herself into places she was not supposed to be. The country was beset by political violence, and in 1984, she had the kind of brush with danger that can make a reporter's name. Pelaez and her cameraman were kidnapped outside their TV station's office in Lima by the revolutionary group Tupac Amaru. They were blindfolded and driven away. The group demanded that Pelaez's TV channel broadcast a propaganda video that was left in a garbage can in return for the pair's release. A few hours after the channel broadcast the clip, in which armed and hooded rebels accused the government of torture, the two were freed. When she returned to the newsroom, she encountered distrust from some of her colleagues. Soon after, she left Peru and came to the United States in 1985 on a visa as a political refugee because she was worried about possible threats from rebels.

She left for America with a new husband. Her first marriage, when she was just seventeen, was to Waldo Marsical and they had a son named after him. But the marriage did not last. It was while on assignment as a newspaper journalist in the early 1980s that she met the man who went by the name of Juan Lazaro. He was a photographer, ten years older but he looked good for it. The two soon became close, as work mingled with her private life. "She was a very passionate woman," a colleague in Peru later said. "To her, he was a hunk." She was a fair bit shorter than him and he taught martial arts. At one point, he pulled her up on his shoulders while out in a story so she could get a look at what was happening at the presidential palace. "I first admired him for his knowledge and ideas of social justice, then I was attracted by his physical strength," she later wrote. On December 3, 1983, they were married. But she had married a lie.

The real Juan Lazaro was a toddler who had died aged three of respiratory failure in 1947, his mother crying whenever she talked about him in the years after. The fake one was another Russian dead double. His real name was Mikhail A. Vasenkov. Little is known of his early life but he is thought to have been born in Moscow in 1942. He left home young, leaving a brother behind whom he would not see again. Nor would he be there when his parents died. He had been selected for the KGB's Directorate S, deep in the Cold War. Once his training was complete, he was sent out on assignment. He had come to Peru, sporting a decent-size mustache, on March 13, 1976, on a Uruguayan passport in the name of Juan Lazaro Fuentes. Spain had been his stop-off point to build the legend. After three months there he had flown from Madrid to Lima, with a forged letter on the stationery of a Spanish tobacco company saying he was coming to the country to carry out a market survey. Two years later he used his passport and a fake birth certificate saying he had been born in Montevideo in Uruguay on September 6, 1943, to request Peruvian citizenship, which he received in 1979. He said he was Uruguayan, but a lot of people found the accent a little odd, more European than Latin American. He rarely spoke of his family.

Lazaro used his cover as a photographer to travel and carry out missions for the KGB. It gave him the ability to meet politicians and businessmen. These were people who might be recruited to become "agents of influence" for the Soviet Union—such an agent did not necessarily provide secrets but instead offered the ability to alter the course of events, small or large, to suit Moscow's needs—perhaps a journalist spreading information or a businessman laundering money or a politician making decisions. Russian reports, whose accuracy is hard to judge, suggest his marriage to Pelaez was genuine (although possibly with KGB approval), and her move to the United States provided an opportunity for him to move his work there, a decision sanctioned by the KGB leadership. FBI officials, though, wonder if Lazaro's target may always have been to go to the United States.

After they moved to the United States, the couple settled in New York and in a house on Clifton Avenue in Yonkers. Pelaez became a US citizen, Lazaro a legal resident. Pelaez resumed her career as a journalist, working first as a reporter and then a columnist for a New York–based Spanish-language newspaper. Her political views were left-wing. She saw herself as standing up to the powerful and speaking for the oppressed and was a critic of American foreign policy, especially in Latin America. In 1993, Juan Jr. was born. He would go on to be a talented pianist who earned a scholarship to a Manhattan arts school. He would play Chopin and Beethoven to audiences, his eyes closed as he seemed absorbed by the music. At a concert in Peru, his proud father was interviewed by an education consultant about how he raised a musical prodigy.

Lazaro was a teaching assistant at the New School in Manhattan from 1993 (eventually earning a doctorate). Later he would be hired as an adjunct professor at Baruch College to teach a class on Latin America and the Caribbean. Like his wife, he seems to have done surprisingly little to hide his politics. Students remember his strident denunciations of US foreign policy. He praised Hugo Chávez, the populist left-wing leader of Venezuela, and attacked the invasion of Iraq as driven by corporate profit seeking.

What did Pelaez know of her husband's identity and spy work? She always maintained she was not a spy and did not know her husband was a Russian illegal. The evidence produced by the FBI, though, suggests she was involved in clandestine behavior. From at least 2000, the FBI was on to the pair. A bugged conversation in the house on February 20, 2002, suggested Pelaez had just returned from Latin America and the couple talked about money she had brought back. A year later, they discussed whether they would have $72,500 or $76,000 after accounting for their expenses following another trip. The bug picked up a conversation on April 17, 2002, in which Lazaro described his childhood to Pelaez. At one point he said, "We moved to Siberia . . . as soon as the war started." If he had been born in Uruguay, why was he brought up in Siberia? It is possible she thought his Latin American communist parents had lived in Russia. Or perhaps, as some FBI officials think, she knew more than she was letting on.

The trips to South America had another purpose as well as collecting money. They were a way of passing covert messages to Russian officials. And they were using one of the most old-fashioned pieces of spy tradecraft. In January 2003, shortly before a trip Pelaez was taking, the bugs in their house captured a conversation between her and Lazaro. He explained he was going to write in "invisible" and she was going to "pass them all of that in a book."

Invisible ink goes back hundreds of years but was still being used by Lazaro to send some of his intelligence reports. The FBI would find pads of papers embedded with specially treated chemicals in their house. An illegal would write a normal-sounding letter to a fictitious friend. Then they would take a sheet of contact paper—almost like carbon paper—and place it over the letter and use a pencil to add a message onto the letter that could not be seen. They would have to carefully destroy the extra papers and mail the letter off to a foreign address—perhaps in Colombia or Austria—or deliver it by hand (as Pelaez seems to have been doing). A Line N officer would receive it and send it to Moscow, where the paper was developed and message decrypted. The whole process was time consuming and slow. But in the digital age, it can be particularly useful since it leaves no electronic trail for investigators to follow.

Communications back to home base are the most difficult and risky part of any spy's work. The whole point of anyone operating undercover is that there should be as little as possible to tie them to their real controllers. If any evidence of contact is discovered, it is highly incriminating. But at the same time, instructions need to flow one way and intelligence back the other. Throughout history there have been many ways spies have sought to manage this process and minimize the risk, from face-to-face meetings to carrier pigeons. In order to preserve their secrecy, illegals were supposed to communicate directly with Moscow Center rather than through officers operating out of the SVR residency in their embassy.

Lazaro was the oldest of the illegals active in the United States. He had been trained in an era long before the internet or digital communications and so used the most old-fashioned techniques—like invisible ink and mailing letters. One counterintelligence official likens illegals to satellites launched out into space. They are sent out with what is state-of-the-art technology at the time, but they then have to keep using that for decades while they operate.

Bringing them back home for training on an entirely new system is not something that can easily be done since it would take so long as to potentially jeopardize their cover. In this way, Juan Lazaro's communications techniques were the most dated of the group, since he had been launched back in the Cold War, pre-digital, pre-internet era. "He was old school," says one FBI officer.

The bug in Pelaez and Lazaro's house also picked up an odd irregular clicking sound on a number of occasions. This, the FBI realized, was linked to the receipt of coded radio messages coming in from Moscow Center. On November 23, 2002, a bug captured Lazaro reading out loud as he composed a lengthy radiogram to Moscow Center about the conflict in Chechnya. Radiograms are coded bursts of data that can be picked up by a commercial radio receiver. This is a classic decades-old communications technique for illegals—still used to this day and which leaves no digital trail.

Twice a week illegals tuned in and then used a one-time pad of seemingly random numbers that were then added to or subtracted from the digits in a message. The papers then had to be disposed of. One option was to fold them tight (like an accordion) and place them on something metal to burn them; another was to soak the paper in water and then rub it with your fingers until it broke up and flush what was left down the toilet. Illegals would spend hours hunched over their notepad, carefully transposing letters and numbers. But all that work was for a reason. As long as only the sender and recipient have the same one-time pad and use it properly, then such a message is impossible to crack even by the most advanced supercomputer. And so, as the investigation began, the FBI's frustration was that it lacked insight into what was being communicated. But eventually they would get their break, thanks to a discovery at the home of another family of illegals. And that would transform the investigation.

8

Breaking and Entering

IT WAS THE middle of the night and the apartment in Hoboken, New Jersey, was dark. Derek Pieper was waiting nervously outside. His job was to be the lookout. The family who lived there in 2005 was on vacation but he had to make sure no nosy neighbors turned up unexpectedly while his colleagues turned over the place. Inside, Maria Ricci and her small team, dressed in black, were hard at work. The team was small because the two-bedroom apartment was. They did not want to literally fall over each other and knock something over. The atmosphere was tense, as it always was. They had to work fast but carefully. Everything would have to be put back exactly where it was. An FBI covert search looks an awful lot like a high-end burglary. But, apart from being legally authorized, the difference was that the aim was not to take anything away that might be missed. The team inside worked their way through the apartment methodically—the couple's clothing, the toys that belonged to the two girls who shared a room. Then they made their way to the TV in the family room. It sat on a large cabinet. In the cabinet were lots of shoe boxes. Inside one were pictures of family trips. The father of the family liked to take photos and there were plenty more like that. Another box had the school report cards for the two girls who lived in the apartment.

The reports were glowing. But one box was different. Inside was a phone, notebooks, and floppy disks. It looked interesting but at the time no one on the team could have known that the contents of that box would transform the investigation into the illegals.

The apartment belonged to the Murphy family. Richard Murphy was supposedly born in Philadelphia. His wife, Cynthia Hopkins, known as Cindy, was from New York. Richard had a round face and was slightly pudgy. He could get grumpy. Cindy was thinner, stylish, with short dark hair. She could appear dour but could turn on a smile when she needed to. They were another pair of illegals, but with different cover than Heathfield and Foley. They arrived in the second half of the 1990s posing as citizens born in America. Doing this had required using fake birth certificates to back up their story. The quality of these certificates, the FBI would later note half-admiringly, was high-class. "They were incredible," says one agent. Using a fake certificate in America relied on the fact that birth certificates were different in each state and varied from year to year, so an official would almost never spot anything amiss unless they were already suspicious. Additionally, there was no central database to check. But even if they were brilliant fakes, they were still fakes, which meant this couple's backstory was weaker than Heathfield and Foley's. When the FBI searched a Manhattan safety deposit box belonging to Richard Murphy, they photographed his birth certificate. They contacted the Philadelphia Bureau of Vital Statistics, who said no record could be found for either Murphy or his supposed father.

Richard Murphy studied international affairs part-time at the New School in New York from 2002 to 2005, a chance to build cover and contacts. In a strange coincidence, his adviser on the faculty was the great-granddaughter of former Soviet leader Nikita Khrushchev. She was puzzled by the young man with an Irish name but a Russian accent. "You know when you meet your countryman even if this countryman speaks a different language and pretends not to be your countryman," she later said. She found it odd that he never tried to speak Russian to her or ask about her family. As a result, she decided she would not ask about his. Perhaps he was someone who wanted to put his Russian past behind him, she thought. "He was a little dour I must say. He was not always happy, which is a bit Russian because, you know, misery is what we do best."

The couple moved into the small Hoboken apartment in 2003 with their two daughters, Katie and Lisa. Neighbors suspected nothing. One family, who also had two young kids, got to know them well. Cindy would cook up a lasagna and cakes—once making cookies shaped like the Statue of Liberty. The two families would barbecue together out on the back patio and go ice-skating and hiking together. Richard and Cindy seemed to enjoy each other's company but were not very "touch-feely," the other couple thought. An FBI team watching their every move would get to know the couple and the state of their marriage much better than the neighbors, though.

The illegals investigation was tightly held within the FBI. It began with the New York field office because that was where Poteyev had been recruited. But once it became clear it was going to offer insights into illegals operating across the country, headquarters in Washington, DC, took on a coordinating function. Initially it was known as "the backroom cases" to the small group read into the investigation in headquarters. It sat with the FBI's SVR unit, whose job was to track SVR activities in the United States. Their main office normally had around six to eight people working together. But there was a small back room that had four more pods that people could work out of. This "back room" was where the illegals cases were coordinated from and where all the highly classified materials were kept. This was a counterintelligence case rather than a counterespionage case. The latter were focused on arresting those, normally Americans, who were passing on secrets to foreign powers; the former were more about understanding the activities of a foreign intelligence service operating inside the United States and more rarely led to arrests.

Alan Kohler, a New Jersey native, joined the SVR unit in 2003, having previously served as a spy-catcher in the Washington field office. A year and a half later he was promoted to chief of the SVR unit and decided it was time to give the whole investigation a formal code name (there were separate code names within each field office and for each target). A computer spits out a list of five options a day for the bureau. If you do not like any, then you can log in the next day and get another five and keep waiting it out until you get one you like. Or you can come up with your own and take it to the team for them to approve. An analyst in the SVR unit came up with a list of names and brought it to Kohler. The others are lost to history, but one stood out as her favorite—Ghost Stories. The FBI tries to shy away from a code name that tells you too much about the case it refers to—after all, the whole point of a code name is that it hides the truth. But agents do sometimes like to come up with something a bit clever or that has an inside joke or reference. In this case everyone agreed that Ghost Stories fit perfectly with the world of dead doubles. And so it was settled.

Ghost Stories would be at once one of the largest but also one of the most sensitive counterintelligence investigations in the bureau's history, revealed in detail here for the first time. The investigation would eventually sprawl across the country, with field offices in Washington, Chicago, Seattle, Boston, as well as New York involved, each passing on news of significant developments through headquarters. The New York field office was the hub for much of the investigative work over the decade. It was where the case had started and many of the illegals would live around the New York region. In 2006, Kohler was promoted to the New York field office to supervise the counterespionage team and became the supervising agent for the illegals cases there, through to the end.

In New York, he would work with two younger officers who would play a key role in the investigation into the Murphys and the wider illegals. Maria Ricci, an Italian-American with an infectious laugh, had grown up in New Jersey. She was not one of those people who had dreamed of being an FBI agent as a child. "If someone said you were going to be an FBI agent, I would have thought it was the most ridiculous thing I'd ever heard in my life," she says. She had been an English major and then trained as a lawyer, but after five years practicing decided she wanted to work in public service. She applied to join the FBI as a lawyer. She was told there were no openings for lawyers but was asked if she would be interested in being an FBI agent. In 2002, to her surprise, she found herself as an agent with her first case, investigating Russian illegals. She would see it through to the end and spend more time on it than any other agent in New York. As a result, colleagues like to joke she was the OG—the "original gangster"—of the team.

Derek Pieper, originally from Boston, was another longstanding member of the team. Quieter, with a dry sense of humor, he had also been to law school after Harvard but decided he did not want to be a lawyer. He had worked as an investigator on public corruption in New York and then joined the bureau's New York field office in 2004. Investigating organized crime seemed the most interesting possibility at the time but instead he found his first assignment was to the counterintelligence squad investigating illegals. He too would see the case through to the end. The squad had to keep things tight even from neighboring counterintelligence teams. "The next squad over had no idea what we were doing," says Pieper. Because the team also hunted for treacherous Americans, it was sometimes assumed by colleagues that their secretiveness might be because they were investigating someone within the building. Working counterintelligence was not the way to make yourself popular.

The FBI agents found themselves plunged into a strange world. Ricci would talk to some of her Boston colleagues who were following Heathfield and Foley about what it took to be an illegal. "This is crazy. I can't imagine going to Quantico and being told: you know what, Maria, you're really good at languages, we think you have an ability to be an illegal—I want you to give up your life, your family, your friends, everything, go to Moscow, live a completely different life. You can't call home anymore. You can't say 'Happy birthday, Mom.' By the way, here's your new husband." They had a front-row seat into the SVR's most secret program and yet the reality was they were watching a family like the Murphys in suburbia. "When you think of a Russian spy you are thinking of someone rappelling off a rooftop or jumping out of a plane," Maria Ricci said years later. "What I think of is these two eating a hamburger at the family picnic and drinking really bad beer."

The FBI's armory of techniques includes monitoring and recording phone calls and emails, searching bank records, placing covert video cameras in public places and hotel rooms, and physical surveillance of suspects. But one of the most important would be the covert searches of their homes. The FBI's job may include combating crime, but one of its core investigative techniques is to do something that, if a passerby happened to spot it, would look a lot like breaking and entering. A covert search is legally authorized to support the FBI's mission. But it still involves people in the middle of the night getting into someone else's property and having a root around. In the case of the illegals this would have seemed even stranger to a passerby. It was not as if these were offices belonging to a company or someone thought to have mob links. These were ordinary suburban houses and apartments belonging to people with families.

Covert searches are risky. If one of the illegals realizes there has been a break-in, either because they notice something in the wrong place or a neighbor tells them there were some strange goings-on when they were out, then they might well suspect the FBI was on to them. In that case, the whole investigation could be over. This meant the searches were carried out only rarely. Typically, they would be done when you knew for sure the inhabitants of a property were out and there was no chance of them coming back. Out for dinner was not good enough—an argument or a bad meal and they could be back early. Out of town for a vacation when you knew the day of their return was better. But even then you still had to worry about neighbors. You do not want them seeing something strange and telling the owners or a police car waking the street with sirens. A property like that of the Murphys would maybe be searched twice a year. No more. "We didn't go in just to have a look around," explains Kohler.

An FBI team wants to know as much as possible about the lifestyle and the property before they go in. The FBI used experts who are "pattern of life analysts," whose job is to learn every detail of people's lives—when do they normally go to sleep and get up? Do they wake in the middle of the night much? When are the neighbors awake? Are there any dogs? What time does the garbage truck come? The last question is important because the FBI carried out what is called "Trash Cover," which means switching trash cans before they are picked up. One source says that for a full decade, the FBI collected all the trash from the illegals' houses to search it for any possible clues. "We owned almost every facet of their life. We knew what they were doing on a daily basis. When they came. When they went," one FBI agent would later say.

A covert entry like that in Hoboken in 2005, which led to

the key breakthrough, would usually take place in the middle of the night. A skeleton key or lock pick got you in. Then windows are carefully blocked out so that the team can use their own light inside without anyone outside becoming suspicious. It would include one or two case officers working directly on the investigation—like Maria Ricci or Derek Pieper—who might be able to spot the significant items. Others would be technical specialists. "You are literally sneaking around somebody's house . . . there's always pressure," says Ricci; "they lived in an apartment building, so walls are thin. So you would hear the guy next door cough and you realize you really needed to be quiet. . . . We went in when we had a reason to go in." Every time you went in, you rolled the dice. Too many times and chances are you would eventually make a mistake. The team searching in Hoboken realized that their targets were using small tricks to detect whether anyone had been inside their apartment while they were away. When the Murphys' closets were opened, they were packed with items. In some cases, strings were weaved around things in a particular pattern. In another there were coins in pockets of certain clothes. Disturbing the string or coins would make it hard to put them back in exactly the same way. They were simple but effective tricks using everyday items. It meant the FBI team decided not to touch certain things, as they could not take the risk.

One role for covert entry teams was the placement of tiny listening devices—microphones or bugs that were able to pick up conversations in the room. This is something the FBI has long experience in developing, so they can be hidden in everyday items. The bureau will not comment on what these might be, where they can be placed, or how they work, for fear of tipping off subjects of investigation. But they gave a deep insight into what targets talked about and how they lived their lives—right down to how they talk to their children and to each other. "I practically lived with the Murphys for so long," says Ricci. "I feel I know Richard Murphy better than some of my relatives, which may say something quite bad about me."

The hours were long for the FBI team, just as they were for the illegals. The amount of material produced by the bugs was enormous. Every time a floorboard squeaked, the recording device would be tripped and the recording would need to be reviewed. The squad of agents had to sit listening to the endless chatter of daily life, hoping somewhere in there was a nugget that might be a clue to some spy activity. The FBI team never heard the Murphys talk to each other about the challenges of living as an illegal. Even when they were alone in the house together, they never broke cover and talked to each other either in Russian or in English about their real work. They certainly seemed aware of the possibility of some kind of surveillance. Occasionally there might be a hushed conversation somewhere in the house that was hard to pick up on the listening devices but seemed to correlate with when there was an operational meeting coming up, but it was hard to be sure.

One of the curious, almost uncomfortable, aspects of the FBI's work was the voyeuristic insight it gave into a couple's marriage and their family life. Living such a strange double life could strengthen the bonds of marriage—providing the sense of a shared mission and the chance to support each other. But it could also introduce strains. The fact the Murphys did not turn to each other to share the burden—even in pillow talk—surprised the FBI team. "It's a little strange that Richard would never turn to Cynthia, but I think there was a competition between them," explains Pieper. The FBI team came to believe that Cynthia felt her husband was not performing as well as he should as a spy and there was tension in the household. Recordings would even have captured the couple's most intimate moments. How do agents deal with that? "Throughout all monitoring, the FBI is diligent in managing what is not relevant," one agent involved in the case explains. "So with regards to intimate moments, professionally you skip ahead."

An important goal in covert searches was to find any evidence of "tradecraft"—the kind of gadgets or paraphernalia that are used by spies and which would provide hard evidence to prove these individuals were Russians. The team would photograph anything that looked even the slightest bit interesting while tech experts would copy any electronic media like CDs or memory sticks and leave them in place. A key question was how the group was communicating back to Moscow Center.

IT WAS DURING the 2005 Hoboken search that the FBI team hit pay dirt thanks to the boxes of photos in the TV cabinet. When the FBI first got onto the trail of the illegals, they could not see inside their covert communications. All they could see was that they were using code pads and encryption techniques that made messages unbreakable. Unlike Lazaro, the Murphys had been in the United States since the 1990s and so were able to update their covert communications to take into account new technology and particularly the emergence of the World Wide Web. From 2000, they started using a new technique, one that it would take the FBI five years to understand. When they did, it would prove to be one of the great breakthroughs in the entire case.

Inside the shoe box were floppy disks and notebooks. This looked interesting, but when you were searching a suspected spy's house, everything looked potentially interesting and you never knew what would turn out to be some kind of hidden piece of spy equipment. But the instincts in this case were right. The box would be pivotal to the investigation. But it would not yield its secrets easily or quickly.

The contents of the box were taken away to be analyzed. An initial forensic computer analysis of the floppy disks found them to be blank. But that did not seem right. So they carried out another check. Again they came up blank. Next the team asked the FBI's computer experts to copy the disks onto other disks—to re-create them so they could play around with them a bit more. Among the papers in the shoe box was one page that had "alt-control-e" written on it along with twenty-seven seemingly random characters. Was it a password? It was time to experiment a little. They put in one of the floppy disks and pressed alt-control-e. Nothing. They tried a different disk. Same process. Nothing. But just when it looked like a dead end, they tried another disk. This time the seemingly empty disk sprang to life. There was a prompt for a password. There was elation for the team. But it was short-lived. They put in the twenty-seven characters from the piece of paper. Nothing. They tried again. Same response. Maybe it had been written down backward? They tried the combination every which way. Nothing. But then another member of the team walked by the agents sitting huddled over the computer. He looked over their shoulder at what they were doing and then at a digit on the piece of paper and said, "You know that's a one, right?" It had looked like a seven because of the way the Murphys had written it, but in Russian it was the way they wrote the number one. "No—we did not know that was a one," they replied sheepishly and quickly went back to the screen. This time as the last character went in, the screen suddenly changed. A prompt appeared. "Please insert picture disc." They were inside the illegals' covert communications system.

Imagine a picture on any normal website. It is made up of data—ones and zeroes—that when downloaded tell a computer how to reconstitute the image on-screen. What if hidden among that data are ones and zeroes that have nothing to do with the image but actually make up a secret message? Thousands of people could visit that website. But only if they had the right software would the message be downloaded. And even if it were spotted by an unintended recipient, it is encrypted, so it cannot be read. This is steganography—the trick of hiding a message inside an image.

Hiding messages has been around for some time. One story from a couple of thousand years ago involves a courier's head being shaved and a message being tattooed on his scalp. The hair is allowed to grow back and anyone intercepting the courier on route will not see anything. But the person whom the courier arrives to see knows that all they have to do is shave his head again to reveal the message. Another example from the twentieth century was the microdot. This was so small as to be invisible and could be put on a stamp or postcard but contained a message that could be read if you knew it was there. By the twenty-first century and the illegals, a new world of digital steganography had arrived.

The FBI team realized you had to insert a disk that had six pictures on it and then another disk with the message you wanted to send. The computer would analyze the pictures to work out which picture was best suited to hiding the message. Once it had done so, it would say something like "number six" to indicate the best one and then encrypt—or scramble—the message. Each time it would be encrypted differently, so you needed a program to be able to decrypt it each time. Moscow Center had created its own bespoke software. In order to extract and then decrypt and read the data, you needed the same SVR-supplied software. Without it the message was unbreakable.

There was also a list of public blog sites on the Web where people could upload their pictures. Richard Murphy loved taking pictures of flowers, and often these pictures would have the message hidden inside them. They would be uploaded to the Web and then at the other end Moscow Center could download and decrypt them. The FBI also copied the hard drive of a computer. On it they found an electronic address book with links to website addresses along with a history of which sites the computer had accessed. They downloaded images from the site that looked entirely innocent—including some of colorful flowers. But when the steganography program was applied, readable text files magically appeared. The FBI team asked their tech expert to create their own version of the encryption program that had every different encryption key on it. They were able to go back and decrypt some earlier messages, as well as new ones that arrived.

The shoe box had been the key to understanding the illegals' communications and would transform the investigation. This item would become known to grateful FBI officers as the "tradecraft box." There was one moment, though, when they feared they might lose their access. One winter Maria Ricci was going back into the Murphys' house just before New Year's Eve. You always did a search for a reason and this time it was because there were indications that there was a change in the communications system—most likely new disks. This was a night search. These are more stressful. Neighbors are likely to be sleeping but if you are discovered there are fewer ways to explain why you are in someone else's house. Ricci was being as quiet as she could with her small team as they hunted for new disks. They found them quickly. But what about the new password to go with them? They were not in the shoe box. Nor any other obvious place. So they started looking anywhere and everywhere else. Still nothing. Minutes turned into hours and the tension was rising. In the end they had to give up. The team made their way back deeply discouraged, assuming they had blown it—without the password they could no longer be able to read the messages. After all that worry when they returned to base and inserted the new disks, it turned out the password had not been changed.

Because of his computer skills, Murphy would later lecture Moscow Center on the technical limitations of the communication system and how hard it was to encode a message. This was gold dust for the FBI, since it helped them further understand the workings of the SVR's top secret system. At one point he explained to Moscow that if the FBI were ever to get hold of the material the illegals had been given, they would have both the instructions and the passwords in one fell swoop. Which of course they had already done. His complaint was absolutely correct, and it was fortunate for the FBI that the SVR did not listen.

When the FBI's Boston team went into the Heathfield and Foley house in 2006, they now knew what to look for. One of the computer disks looked similar to the New Jersey find and it too required a twenty-seven-character password. Traces of deleted electronic messages were found that FBI technicians were able to recover. These were drafts of messages sent to Moscow using steganography. Other illegals would also use the technique. The break into the communications was critical for the case. "For us, that was a gamechanger," Tony Rogers of the Boston FBI field office later said.

The FBI could not only read the messages going back and forth between the illegals and Moscow Center but—thanks to the bugs in the house—they could even hear the illegals sometimes discussing what they thought of what Moscow Center was telling them to do and how it made them feel. That was something even their SVR controllers would not know. "That's everything," says Alan Kohler. "There's nothing going on with this cell that we don't know about."

This allowed the FBI to stay one step ahead of the illegals. If they were planning a covert meeting, then there was no need to follow the spies to find out where it was and risk being spotted. Since you knew exactly where and when the meeting would take place, you could simply stake out the location ready for them to arrive. But even more important, it provided an insight into what orders they were being sent from Moscow and what intelligence they were sending back. Their overall mission was set out in a message sent to the Murphys: "The only goal and task of our Service and of us is security of our country. All our activities are subjected to this goal. Only for reaching this goal you were dispatched to US, settled down there, gained legal status and were expected to start striking up usefull [*sic*] acquaintances, broadening circle of your well placed connections, gaining information and eventually recruiting sources."

This was the mission of the illegals. To pose as Americans, bury themselves deep, and then identify people who could help Russia. And in Moscow, there was a new master for Russia's spies. He was one of their own and a man for whom the importance of spying—and catching your enemy's spies—was utmost in his mind.

9

Putin's Spy Fever

On December 20, 1999, hundreds of Russia's spies took refuge from the bitter cold outside as they gathered inside the Lubyanka and Yasenevo, the headquarters of the FSB and SVR. They were there to celebrate the anniversary of the founding of the Cheka—Russia's revolutionary secret service—which for convenience's sake had also been made the birthday of both new services. Champagne glasses were raised in a toast to Felix Dzerzhinsky, the man whose statue had been toppled outside just eight years earlier. The birthday party took place every year. But this time it was different. It had been a hard decade for the spies, as they lost their place as the elite after the end of the Soviet Union and as their old adversary seemed to take the upper hand. But the atmosphere that night was more optimistic than many could remember. In the Lubyanka, the special guest was the country's new prime minister. Only a few months earlier he had been running the FSB. "Dear Comrades," Vladimir Putin began, "I would like to announce to you that the group of FSB agents that you sent to work undercover in the government has accomplished the first part of its mission." It was meant as a joke. But it was also true. "There are no former agents," Putin told the crowd, adding another twist to the old joke about there being no such thing as an ex-KGB officer. A veteran of the KGB and FSB was now Russia's leader. Russia's new prime minister had risen from nowhere with a promise to lay to rest the ghosts of a turbulent past decade.

Putin had risen out of the chaos of the 1990s and it defined him. He would describe the end of the Soviet Union as "the greatest geopolitical catastrophe of the twentieth century." This was because of what he saw had come after—a weakened Russia and a more dangerous world with no one to balance American power. Growing up a KGB officer in the dying days of the Soviet Union had given him a conspiratorial view of the world in which Russia's enemies had constantly been fueling division to keep his country down. A great country had been humiliated. Yeltsin had appeared drunk when meeting President Bill Clinton, another source of shame. Only a former KGB man, in his eyes, could restore the strength of the state, which in turn protected the motherland. Restoring Russia would require dealing with two groups who had exploited its weakness—oligarchs and foreign spies. First to be dealt with was an old mentor, an oligarch whose power was closely linked to Putin's rise.

When Alexander Litvinenko had returned to Moscow after his time in Chechnya, he found the FSB embroiled in not just corruption but murder, carrying out vendettas for politicians and criminals. Working in the economic crime directorate, he was now asked to look into assassinating the man he had once kept an eye on—the deputy head of the national security council, Boris Berezovsky. The oligarch had negotiated a peace deal with Chechen rebels that hard-liners perceived as a sellout. Litvinenko instead told Berezovsky about the plot. As a result, the then head of the FSB was sacked. His replacement in the summer of 1998 was a surprise—a midranking blank-faced colonel named Vladimir Putin. Berezovsky had known Putin since the start of the 1990s and even vacationed with him. He thought Putin would be his man and he arranged a meeting for Litvinenko with his new boss.

As Litvinenko entered Putin's office in the Lubyanka, the new FSB director came out from behind his desk to greet him. There were no pleasantries. "We operatives have a special style of behaviour," Litvinenko later wrote. "Just look into each other's eyes and it becomes clear, do you trust the person or not. And I immediately had the impression that he is not sincere." Putin was not a tall man and took up martial arts at school so he could stand up for himself. He liked to intimidate. The two spoke for forty minutes. Litvinenko outlined his knowledge of corruption in the FSB and its links to organized crime. But when Litvinenko offered to hand over his written summary, Putin declined. "You keep it, it's your work," he told him. Litvinenko later claimed Putin invited him to join his "team" but he refused. As soon as he left the room, Putin ordered an investigation against him.

A few days later, on November 17, 1998, Litvinenko went public in his criticism of the FSB. He and a group of colleagues organized a dramatic press conference. The other FSB officers wore masks to hide their faces but Litvinenko did not. The media attention was huge but hopes that it would lead to change were dashed. Instead it marked Litvinenko out as a man who had betrayed his colleagues in the eyes of Putin. The FSB head gave some public remarks soon after and singled out Litvinenko in a strangely personal way, talking about his marriage, among other things. This was a sign of just how personally Putin took betrayal. He was a man who kept grudges. Litvinenko was arrested in March 1999. When he was acquitted in his first trial, FSB officers stormed into the courtroom and arrested him again.

Yeltsin was now ill and his priority was to find a way out that protected his family from corruption charges. He and his allies, like Berezovsky, needed an empty vessel they could fill. FSB director Putin was their choice. After all, a former KGB man was perfect thanks to the cult of the spy built up over the decades. He was made prime minister in August 1999. A series of devastating bomb blasts hit apartment buildings in Moscow in September. It seemed to confirm the need for a strong man to take on the Chechen insurgents paving the way for a new war. Yeltsin resigned suddenly on New Year's Eve in 1999 and Putin—from almost nowhere—was acting president. He would be formally elected in March. He was chosen because he was thought to be a blank slate without his own politics. But the vessel was not as empty as it looked. This former spy had his own views. Russia had been humiliated. Now it was time to push back.

In July 2000, Putin summoned the oligarchs to a barbecue and delivered a blunt message—if they wanted to survive they would have to bend the knee. They would serve the state and not the other way around. Stay out of politics and you could keep your wealth. Or face the consequences.

Berezovsky soon learned he had underestimated the man he had sponsored. His TV stations criticized the new president when the *Kursk* submarine sank, with the loss of 118 lives, in August 2000. Putin was lacerated for vacationing while the sailors perished. Putin summoned Berezovsky to the Kremlin and told him to hand over his TV channel. Berezovsky refused. He fled to the United Kingdom and claimed asylum. From there he began to plot. He soon helped Litvinenko also flee to London. Once there, the former FSB officer received a phone call saying "Remember Trotsky"—the exile killed by an assassin sent by Moscow. Back in Russia, the oligarchs would fall into two categories. Those who accepted the new reality of Putin and the state as their master and those who tried to fight it. The latter ended up in jail, in exile, or dead.

Spies—the threat of foreign ones and successes of Russia's

own—would be a defining theme for Vladimir Putin. He may not have been a first-rank KGB officer in his own time (which may have only added to his infatuation with the world of espionage), but he both believed in the cult of the spy and understood its power among the public. The new leader breathed life into his own decaying spy services, turbocharging them with more resources and a renewed sense of purpose. Just as he would build a cult of personality around himself, so Putin would build one around his spies. They would once again become heroes and the source of pride—and that particularly applied to illegals. Russia may not be an economic giant, but one area where it was still a first-class power was espionage, and Putin would double down on his intelligence services as a means to wield power and influence around the world. But equally important was the focus on the villainous, subversive work of enemy spies operating in Russia and those treacherous individuals who agreed to work with them. In May 1999, when still running the FSB, Putin in a newspaper interview had identified foreign espionage as the nation's biggest threat. Those spies were not just using diplomatic cover, he warned, but also using businesses, charities, even ecological organizations. "Spy fever" would be introduced into the body politic and periodically stoked up. It would be a powerful tool for establishing a pervasive sense of threat—that Western countries and particularly the United States and United Kingdom were intent on undermining Russia and preventing her from returning to her rightful position. And, of course, the complaints about foreign spies were sometimes true. Western spy services had been busy. Just because you are paranoid, it does not mean they are not out to get you. The unrelenting Western espionage campaign had fed the paranoia of the Russian leadership and now would provide a justification for a crackdown.

Spy fever came in bouts, with different symptoms each time.

As Putin moved from FSB to the premiership, there was a particular fear that Western secret services were stealing Russian defense technology—the one area where the level of sophistication rivaled that of the West. The openness of the 1990s had created all sorts of contacts and partnerships that could be exploited. One case would yield another of the four men who would be swapped for the illegals in 2010.

A few weeks after Putin became prime minister, the FSB burst into the apartment of a Russian scientist named Igor Sutyagin. The trail led back to a conference in Birmingham University on British-Russian relations the previous year. The professor organizing the event had received a fax from Alternative Futures, a London-based political risk consultancy, asking if they could send someone to attend. Three days later the professor received a telephone call asking for a Russian researcher to be invited. The researcher was Sutyagin. Born in 1965, he had studied physics at Moscow State University and ended up working at the US-Canada Institute, a think tank, as an expert on nuclear weapons and arms control. At the conference, the professor thought Sutyagin seemed pleasant but low-key. "He was small, slight and mild-mannered but aggressive in argument as academics tend to be." But the man from Alternative Futures, who went by the name of Sean Kidd, was different. "Mr. Kidd was a bit of a flashy dresser who drove a sports car. He didn't talk much but he took a lot of notes," the professor recalled.

A year and a half later, Sutyagin was arrested. It was alleged that in Birmingham and then during a visit to London he was recruited by US military intelligence. The initial charges were so vague that a regional court said they were "impossible to understand." But the case was simply returned to the FSB to do better next time. Sutyagin was accused of having met with Alternative Futures in the United Kingdom, Belgium, Italy, Poland, and

Hungary and having passed on information about, among other things, air-to-air missiles, the MiG-29 fighter jet, and Russia's strategic nuclear forces. The list sounded sensitive. But there was a problem. Sutyagin did not have access to any secrets. He was a researcher in a think tank. Sutyagin was adamant: he was no spy. He did not have a security clearance and had only passed information that was publicly available, believing it was for a business consultancy. He explained that a report on an early warning system was based on information in the *Washington Post*. Another on the creation of new military units was from public statements by Defense Ministry officials.

What was Alternative Futures? Sean Kidd had told Sutyagin it was helping businesses in the West understand the investment climate in Russia and would pay $1,000 a month for "exclusive" information. That was serious money since Sutyagin was only paid about $100 a month by the institute. Russian investigative journalists found Alternative Futures had vanished without a trace. "I wish these people would just come and tell their story," the head of Sutyagin's institute said of the company. "It sounds like something from John Le Carré. Are they just going to wash their hands of it and see a young man destroyed?"

The FSB was right that Alternative Futures was a front for Western intelligence but was wrong about whom for. It would later emerge that Britain's MI6 rather than US military intelligence was behind it. Western spies would often try to exploit the knowledge that in the hard-up Russia of the 1990s, foreign money was sometimes what allowed researchers to feed their families. Sutyagin always said he did not realize what was happening and there is no evidence he became a spy.

On April 7, 2004, he was found guilty. Sentenced to fifteen years, Sutyagin disappeared on a grim tour of the Russian penal colony system, ending up in the northern city of Arkhangelsk. Sutyagin's case became a cause célèbre for human rights groups who campaigned for his release, taking it up as an example of an increasingly authoritarian state criminalizing contact with foreigners. In MI6, there was a sense of guilt over a man languishing in a prison cell.

Two days after Sutyagin was pronounced guilty, a high-ranking Russian spy spoke anonymously to the media. The story's headline was "Spy Case Shows That Russia Is Recovering—Secret Service Source." The Russian security state believed that everything had been for sale in the Russia of the 1990s. Now a message was being sent. Times were changing. It put Sutyagin's case into a wider context of British and American spying. "Over the last five years foreign intelligence services have considerably stepped up their efforts," the official said. They argued that the CIA's priority had been political interference rather than just gathering information. "The CIA had the task of controlling political events in Russia and directing those processes into a route desirable for Washington," the source said. This was the new mantra—foreign subversion, aided by domestic traitors. Russia's spy-catchers like Zhomov were on the hunt—determined to erase the humiliation of the past. Their reach would extend from the streets of Moscow all the way to the suburbs of America.

Alexander Zaporozhsky had cashed in the $2 million he earned from the tip-off pointing to Aldrich Ames and then retired from the SVR, content that no one knew his secret. He moved to America and was resettled with the help of the CIA. A few neighbors in the Maryland suburbs were curious about the man with the thick accent who had moved into the million-dollar mansion in a gated community. His explanation that he worked in some kind of vague "import-export" business only added to the mystery. "My guess was he was in the porn business," one neighbor said later. "How else can you make dough like that?" Like many of his former colleagues, he figured he could cash in on his connections and offer consultancy to those seeking to do business between America and Russia. He traveled to Europe and would occasionally drop in on former colleagues in Russian embassies.

In 2000 he met two of his CIA handlers at a restaurant for one of their regular lunches. "I'm going back to Moscow for a visit," he told them, explaining he had been invited back to some kind of reunion with old KGB friends. "That's not a good idea," they told him before spending the rest of lunch trying to dissuade him. He waggled his finger at them and said they were wrong, explaining he had already been back once to Moscow, in the summer of 1999, without any problems and he had been reassured by former colleagues that there would be no issues this time. He did not say which former colleagues had invited him, but the Americans later learned that among them was a woman in whom he had more than a passing interest. They again said no, you should not do this. Concern escalated up the agency. "He can't do that," Steve Kappes, the head of counterintelligence, said when told about the plan, leading him to go and see the Russian himself and explain it was too dangerous. Zaporozhsky was even told that CIA director George Tenet did not want him to go. But the strong-willed spy was now an American citizen and there was no way of stopping him. Zaporozhsky thought he was smarter than everyone else. He thought he could still play the game and win. And he suffered from the spies' folly, the belief they will never get caught. "This is probably the last time I'm ever going to see you," said one American as the Russian prepared to head back. As soon as Zaporozhsky got off the plane in Moscow in November 2000, the FSB was waiting. The handcuffs went straight on. He would be sentenced to eighteen years' hard labor. It was a trap and the CIA's old foe Sasha Zhomov had masterminded the operation.

The reason the Americans had been so keen to dissuade Zaporozhsky from returning to Russia is that they knew there was another traitor who might compromise him. Robert Hanssen was responsible, in different ways, for the fate of two men in the 2010 spy swap. He had been able to search FBI computer systems and monitor investigations into spies. One search came back with information that he passed to the KGB in a dead drop. This had led them to Zaporozhsky and his arrest. Back in November 1987, Hanssen had passed another cable to Moscow. This one reported a meeting between a CIA and a KGB officer, another of the four swapped in Vienna. Jack Platt, known as "Cowboy" because of his boots and devil-may-care attitude, had been assigned by the CIA to target Soviet diplomats in Washington. He had struck up an unusual friendship with one of his targets—Gennady Vasilenko of the KGB. The pair met at a Harlem Globetrotters game and got along. Platt took the Russian hunting in the woods and got drunk with him over long lunches. This was not the first time a spy had told his bosses that he just needed one more boozy lunch to recruit his source. But neither recruited the other. After he left Washington, Vasilenko resurfaced in Guyana. Platt flew to meet him and cabled the results of their conversation back to the CIA, with the FBI copied in. Just days after that meeting, Hanssen handed over the cable to the KGB. Vasilenko had not told his bosses about all of his contacts and so was shipped back to Moscow on a freighter. Since there was no proof he was working for the Americans, he was eventually released.

The hunt for a mole had continued in Washington in the late 1990s after the arrest of Aldrich Ames. The FBI zeroed in on a CIA officer (who was the agency's leading expert on illegals). His home was searched, his phone tapped, his daughter (a CIA employee) was told her dad might be a traitor. He was not. The FBI also came up with a plan that involved identifying Russians who might know the identity of the mole and "cold-pitching" them—walking up to them, often when they were traveling abroad, and offering them a million dollars. Their list had two hundred names.

In 1999, Vasilenko ran into a retired former KGB colleague in Moscow who was in trouble with the Russian mafia and needed money. Vasilenko introduced him to Platt, who in turn passed on the name to the FBI team hunting the mole. They invited him for a spurious meeting with a potential business partner in New York. Vasilenko signed documents for the man's travel. In April 2000, number 28 of the FBI's million-dollar pitches hit the jackpot. The former KGB man revealed he had hard evidence regarding the American traitor. Hanssen had taken one smart precaution. He had never told his Russian handlers his real name. He had left documents and instructions in dead drops and picked up the money in return. The KGB had never liked running an agent whose identity they did not know so they had collected evidence, including a plastic bag he had used at one of the dead drops and the recording of a telephone conversation between the mole and a KGB officer. The former KGB officer had stolen the file containing this physical evidence and kept it at his mother's house as insurance for a rainy day. Now he cashed it in for a $7 million payoff. When they played the tape recording, FBI agents expected to hear the voice of the CIA officer they were convinced was the traitor but instead listened in shock as they realized it was their own colleague Robert Hanssen. One agent threw his headphones against the wall.

Hanssen had handed over everything, from the names of agents to details of hugely expensive technical surveillance programs. This included Operation Monopoly. A house had been purchased in Washington, DC, to allow a tunnel to be built under the Soviet embassy in Washington to facilitate eavesdropping. The total estimated cost of Hanssen's betrayal was in the tens of billions of dollars. He had even given the KGB the inside track on how the FBI tried to catch them on American soil. "Hanssen had destroyed the nation's defenses against Russian espionage," writes Mike Sulick of the CIA. "With knowledge of the FBI's physical and technical surveillance, the Russians could operate unhindered on American soil." But fortunately, he had not known everything. He did not know that the FBI had recruited Poteyev in New York the previous year. On February 18, 2001, Hanssen was arrested as he went to a drop in the woods. Ames and Hanssen, along with two other spies, explain most of the compromises of the 1980s and 1990s. But not all. Multiple CIA and FBI officers say they continue to believe there was at least one other mole who operated during that time and who has never been unmasked.

Inside the FSB there was fury at the loss of a prized source who had spied for them for decades. They wanted revenge. And the man tasked with exacting it was, once again, Sasha Zhomov. In 2003, Milt Bearden, now retired from the CIA, sat across a table from Zhomov. "I can still see Zhomov's steel-gray eyes darken as we sat across from each other in Moscow, the Russian counter-intelligence officer fingering a delicate silver crucifix on a chain around his neck and promising me that he would not rest until he had found the traitors in his midst," Bearden wrote later. The man who had sold the file was out of reach, but Zhomov would find others to hold responsible. In 2005, Vasilenko was arrested again. His signature had been on the travel documents of the man who had given Hanssen away. He was interrogated personally by Zhomov, although the FSB officer would leave the rough stuff to others. In prison Vasilenko met his old KGB colleague Zaporozshky. One had foolishly left America. The other now felt sold out by his friend over there. Together, they faced years of hard labor. The Russians were no longer on the back foot. Under Putin, they were determined to hit back after the disastrous nineties. And that meant increasing their focus on catching those spying within Russia while also driving forward the work of their own spies operating in the West.

10

Targeting

On December 3, 2004, Donald Heathfield composed a secret message for his superiors at Moscow Center. Their deep-cover spy was beginning to make inroads in his primary mission. He told the SVR that he had just attended a seminar. He had managed to strike up a conversation with a serving official at the Department of Energy who worked for a US government research facility that carried out strategic planning for nuclear weapons development. Heathfield reported that he had conversations with the official about research programs on small-yield, high-penetration nuclear warheads (a type of so-called bunker buster). This was promising. Soon after, in September 2005, he would compose another message. Now he had "established contact" with a former high-ranking US national security official. The primary mission of the SVR is to recruit agents who can provide useful information. And one of the jobs of the American-based illegals was to scout these targets.

Heathfield's work was superb cover. With multiple degrees, including from Harvard, the days of selling diapers were now long gone. He first worked for Global Partners Inc., a business development consultancy that offered strategic planning and that opened doors in business and government. Colleagues remember a well-organized, hard worker. His starting salary was roughly $85,000—paying the bills was getting easier. He would claim his clients included major companies like General Electric, Ericsson, Motorola, Microsoft, and T-Mobile, although it is not entirely clear how far this really stacked up. But he was also building a wider network of contacts.

"He'd always hang around and always want to meet people," says William Halal, a professor at George Washington University who got to know Heathfield after he started to turn up at events. The spy began to hang around with a crowd who shared an interest in strategy, including those at think tanks, universities, and federal agencies. Halal was an expert on technological change and worked with companies but also government agencies, including the Pentagon, to predict the impact of emerging technologies. "He wanted to collaborate—that's what drew me in," Halal recalls, and the two began to cooperate. "He wasn't fooling around. That's why he carried it off so well." Others recall Heathfield's interest in how America would adapt to long-term changes in energy supply—an issue of major interest for Russia, whose economy depends on oil and gas exports.

In 2006, Heathfield struck out and founded his own company, called Future Map. It claimed to have offices not just in Cambridge but also Paris and Singapore. It offered something rather nebulous—help for companies to plan for future challenges. He shared the trait of many management consultants with a preference for vague business-speak, which made it hard to pin down exactly what he meant. Along with talk about "developing strategic proactivity," he offered a five-thousand-dollar software package to help CEOs keep track of long-term plans and spot possible surprises. "He was basically using what amounted to a calendar with a few bells and whistles," an intern who was asked by Heathfield to help develop the software said. "It was a piece of junk." But it was the perfect product to pitch to interesting people.

Heathfield tried to market his software to government agencies, often through intermediaries. It was not just a sell-and-forget piece of software, but had the option of an ongoing relationship. He met an employee of STRATFOR, a private intelligence and forecasting company, five times to pitch to them. The company's founder later said he feared if they had installed the software it would have secretly sent data to Moscow, but FBI officials do not believe there was anything hidden inside it (although perhaps that could have been added down the road). FBI agents recall listening in to an endless number of Heathfield's sales calls. He did not quite have the smooth patter of an American salesperson to close the deal, they felt. There was always something a little abrupt when it came to the final "are you going to buy this or what" (they speculated as to whether this was peculiarly Russian). They thought he was a little arrogant, but FBI agents also thought he was the "truest believer" of the illegals.

The electronic messages the FBI intercepted would be the key to understanding what the illegals were up to. They made it clear that they "spent a great deal of time collecting information and passing it to Moscow Center," according to the FBI. However, little of this has been made public. That is partly because these messages included the names of hundreds of Americans—each of them a potential target for Russian intelligence.

"You have got to have more contacts. That was their mantra over the years," one person who served as an illegal in the West recalls of his orders. Heathfield, like the other illegals, sent back hundreds of names of people they met. Each possible target was given a code name. Heathfield had met one person designated "Parrot." "Your relationship with 'Parrot' looks very promising as a valid source of info from US power circles," the SVR said. "To start working on him professionally we need all available details on his background, current position, habits, contacts, opportunities, etc." Once this was sent back to Moscow they would research him carefully—this might involve looking for any personal "weaknesses" like debts or problematic relationships that could be exploited. But it could be something more mundane about his interests that would allow Heathfield to build a relationship and get close to him. These two initial steps—of identifying someone who might be a potential agent and then learning about them—are known in the trade as "spotting" and "assessing."

Intelligence agencies like the SVR build up vast libraries of files on individuals who might end up one day being of interest. Much of this material will not be "classified" but will be everything from university history to gossip to their favorite sports team. You never knew which detail might end up being useful. When that person needed to be approached, you might be able to drop into the conversation the name of your favorite team. Perhaps they should meet up for a beer before the game next month?

One thing that the illegals were not normally supposed to do was the final pitch to someone to become a spy—the "recruitment." Showing their hand as SVR officers was too risky. Most pitches do not work. And if an illegal pitches and fails and it gets reported, then all the investment in their cover is blown. The FBI sometimes saw the illegals champing at the bit. "I want to go after this guy," they would be saying, but Moscow Center would say no. The SVR would frequently remind the illegals to remember the security rules that had been laid out. This included trying to keep the deep-cover illegals away from getting too close to anyone with a security clearance. If that person put down the illegal as a friend on their clearance form, then their cover might not stand up to an investigation.

The illegals supplied material for other SVR officers to make a pitch—creating what are called "targeting packages." In the long term, some of the leads they sent back to Moscow may have proven fruitful or may do so in the future. Intelligence recruitment is a long game that is played out over years and even decades, especially by the Russians. Perhaps some American will travel to Moscow in a few years' time and a file will be pulled with a personal detail about their particular interest or vice that was collected by Donald Heathfield two decades earlier and that detail will allow the SVR to find just the right way of approaching and turning him or her. We may never know.

How successful was Heathfield in talent spotting? In one message, Moscow Center told Heathfield: "Agree with you [*sic*] proposal to keep relations with 'Cat' [a former high-ranking US national security official he met in September 2005] but watch him." That individual is thought to be Leon Fuerth, who had been national security adviser to Vice President Al Gore. Heathfield introduced himself after Fuerth gave a speech and went on to propose they become partners in a research project on long-range projections. Heathfield made a point of citing Fuerth multiple times in the introduction and body of his paper on improving decision making in national security. But Fuerth declined to get involved with Heathfield, who soon after gave up trying to communicate. The illegals did get close to government officials—at least one official was approached to confirm their identity by the FBI. But their success was limited. There was a reason for that. The FBI was watching closely.

The FBI observed Heathfield report back on the contacts he cultivated. To their amusement, they could see times when he exaggerated his success (something spies through the ages have been guilty of). But a reason Heathfield may not have been as successful as he would have liked was that if the FBI did see him getting too close to someone, they would intervene. They will not say how often that happened, but it seems to have been on multiple occasions. It was not always a case of confronting someone with the truth that their new friend was a Russian spy, as that carried risks of their investigation leaking out. It could be easier to just make sure people's paths did not cross. "If they get too close to someone we don't want them to get too close to, we just move them," explains FBI agent Derek Pieper. "Fundamentally that was our job. It was to make sure they never got access to classified information. It is constantly that balancing act. Are they getting too close—how do we stop them—what do we do—without blowing the case." But their success in keeping the illegals away from classified material itself created a challenge for the FBI team. Colleagues in the bureau would ask if the illegals had recruited anyone with access to classified information. The answer would be no. So what's the big deal, they would ask? Why spend so much time and resources on watching them? The team would have to explain that the threat was much longer term and subtle. What if the illegals spotted someone who could be recruited and then ten years down the line wound up working as a penetration agent inside the American intelligence community? After all, when Kim Philby was recruited as a young student fresh out of Cambridge, he had zero access to classified intelligence, but a decade and a half later he was the liaison between MI6 and the CIA and providing the most sensitive of secrets.

The risk with the illegals was not that they would burrow into government or intelligence agencies themselves. Rather, it was that they would be able to recruit people who could do so in the long term. This was what made the generation known as the "great illegals" in the 1930s such a success. In Britain, they had talent-spotted a group of young men fresh out of Cambridge University because they had seen their potential. The "Cambridge Five" (although there may have been more than five and an Oxford ring as well) were directed to work their way into the heart of the British establishment—which they managed in the following years, getting into MI5, the Foreign Office, and, in the form of Kim Philby, the higher echelons of MI6. The goal of Moscow had always been to replicate this kind of success. "We believe the SVR illegals may well have hoped to do the same thing here," said an FBI counterintelligence agent. "They identified colleagues, friends, and others who might be vulnerable targets, and it is possible they were seeking to co-opt people they encountered in the academic environment who might one day hold positions of power and influence." This was the real danger posed by their work: that they might recruit a new generation of spies who could themselves access secrets. "It is ultimately the long game," explains Maria Ricci. "It is not who you know today—it is what that person you know today becomes in ten years."

Boston, and especially Cambridge—home to Harvard—was an ideal location, full of thought leaders, policy makers, scientists, and politicians. Heathfield and Foley lived a good life with nice cars and foreign travel. They told neighbors they were French Canadian, although one—who happened to be a French teacher—found the accent a little odd. Ann looked after the kids when they were young. While the FBI found Donald to be quite stern, she struck them as a good mother, conscientious with her kids. There were barbecues and baseball on the weekend. As they grew older, she worked in real estate, latterly for a company called Redfin, based in Somerville. "She was nice, friendly, very normal. Isn't that what they always say about the guys next door who turn out to be Russian spies?" her boss later said. The couple had a division of labor. Ann—at home more—would handle most of the communications back to Moscow, while Donald did the spying. The FBI would overhear the odd arguments over the timeliness of the sending of reports back, but she was capable of standing up for herself.

FOR THEIR TWO children, Alex and Timothy, the only distinctive aspect of their upbringing was how international it was. Timothy had been born in June 1990 and spent his first five years in Toronto, then France, before coming to the United States. Their parents deliberately chose an international, bilingual English and French school for the boys. "We were trying to avoid making them typical Americans," their mother later said. One thing their parents could not countenance was the idea that children in American schools had to give a pledge of allegiance every morning. Perhaps there were just too many echoes of the pledge that they had given to their motherland before they set out as illegals. There was also a desire to broaden their outlook. They understood that perhaps one day their boys might learn that their lives had been built on a lie and would have to adapt. "We tried to have as many opportunities as possible to see and compare different countries," Heathfield later said. "It is obvious that living in another country, you cannot join the Russian values. But you can instill—if not love, because they do not know the country—then at least respect." The children remember their father encouraging them to read, travel, and be interested in the world. "As a family we loved to travel and did so extensively when we could," Alex later recalled. They went to about forty countries. Their mother never cooked them borsch, a Russian dish, but she did once make them *pelmeni,* another dish from back home, and claimed it was Italian tortellini, a small deception. When Alex was struggling with math he was sent for extra lessons at a school founded by Russians. He began telling his parents how good Russians seemed to be at math. He had no idea that, at least in terms of his own ancestry if not upbringing, he was himself Russian. In the back of the parents' minds there was always the question of what the boys would find out and when.

BESIDES THEIR LONG-TERM goal of spotting possible agents, the illegals were also given specific intelligence-gathering directives from Moscow Center—so-called info tasks. In April 2006, an electronic message provided instructions for the coming two months. This included gathering information on US policy on an eclectic range of issues—from terrorist use of the internet to policy in Central Asia and US views of Russian foreign policy. Heathfield and Foley dutifully compiled responses. In May they sent a message about the arrival of Michael Hayden as the new CIA director. They also sent details about the upcoming 2008 presidential election. This information was described as having been received in a "private conversation" with a former legislative counsel in Congress who was now a member of an economics faculty at a university but who continued to have contacts within Congress and policy-making circles. Espionage was not like a crime in which you were hurting someone else, Heathfield would argue. Rather, it was a patriotic activity, a defensive exercise of gathering intelligence to help your country's leaders make the right decisions. This, he would say, could almost be a "stabilizing factor" in international relations that allowed leaders to cut through propaganda and misinformation by seeing the true picture. Predicting US foreign policy toward Russia was one of their primary missions. "You have to foretell the intentions of the people who build up political plans," Foley later said. "It is important to anticipate what decision will be taken and pass over this information in time for our country's leadership to get ready for it."

Heathfield never thought of the Americans around him as enemies. He considered himself more like an undercover anthropologist on a mission to understand. If you behaved like James Bond, he said, you would survive half a day. If you thought there was some safe where all the secrets you needed were hidden and you just needed to break in, then you were wrong. The goal was more subtle than stealing secrets. "The highest class of intelligence is to understand what your opponent will be thinking about tomorrow, and not what he was thinking about yesterday," he argued.

Putin and those around him came from the secret world. That meant they valued secret information particularly highly. They often only believed a piece of information if it came from a secret source rather than something public. This was part of why the illegals were valued in Moscow. Even if their intelligence was not always revelatory, it was given extra weight because of how it had been obtained. One other important lesson for those trying to understand the Kremlin was how little Putin—who had barely traveled abroad apart from East Germany before he came to office—really understood the West and particularly America.

Richard Murphy proved less successful in his mission than Heathfield. He had originally been the primary operative—the person who was supposed to go out and get a job and develop useful contacts while his wife supported. He was a computer technician and in the early 2000s got a job at the G-7 Group, which advised its clients on how government policy affected financial markets. It included influential figures like a former Federal Reserve vice chairman and fund-raisers for political parties. It was a good place to start building contacts but Murphy only lasted three years because he did not have the technical skills the firm required. His struggles as a spy soon became a source of tension for the couple.

On September 23, 2004, a bug inside their Hoboken apartment picked up a conversation in which Cynthia lectured Richard. Sometimes a partner might lecture their spouse that they need to get a promotion or a pay raise, but she was telling him he needed to be a better spy and raise his game. The problem, she explained, was that he would not be able to get a job in the upper echelons of the US government, for instance in the State Department. This was because of their birth certificates. Any proper background check risked exposing the certificates as fake. So instead, she told him, he had to do more to approach people with access to places like the White House to gain indirect access. So far he was not getting close.

Soon Cynthia made clear that since she did not think Richard was up to it, she was going to take over. So the couple swapped roles. She would go to work and do the intelligence gathering and he would look after the kids. It did not take long for it to become clear that Cynthia was the more capable spy. She had studied finance and international business at the Stern School at New York University. Her LinkedIn profile says she had worked since 1997 as a vice president at Morea Financial Services, which provided "comprehensive Financial Planning and Tax Accounting services for high net-worth individuals"—perfect cover to meet influential people. Later, between 2008 and 2010, she would study for an MBA at Columbia. This proved an even better hunting ground for her. She, along with Heathfield, would be the most successful of the deep-cover illegals.

What is it like to sacrifice a normal life in order to be an illegal—for the adventure of being a spy—and then find out you are not very good at it, and your wife is more talented? Rather than being out meeting people and cultivating contacts, Murphy found himself stuck at home. This loss of purpose left Richard Murphy struggling and unhappy, the FBI thought as they watched. "I think Richard Murphy was clinically depressed," says Alan Kohler. "He just sat on the couch and watched TV all day long and Cynthia was like screw this—we are here for a job—I'm going to do the job. So then she started doing it." The FBI would sometimes hear them argue. Cynthia would call from work and ask if he could help with something—she was busy, could Richard pick up the kids? He would mutter something about being busy doing the vacuuming. "And then we would hear him watching *The Sopranos*," said Pieper.

Juan Lazaro was also having problems with the intelligence he was sending back. On September 10, 2002, he and Pelaez were overheard discussing Moscow's unhappiness with Lazaro's recent reporting. "They tell me that my information is of no value because I didn't provide any source," he complained despondently. "Put down any politician," Pelaez says, apparently happy to pull the wool over the eyes of Moscow. It sounds like he then complains about his bosses not caring about the mission. "So . . . why do they have you? If they don't care about the country . . . What do we have intelligence services for?" Pelaez replies.

Lazaro's career was on the wane and he retired in 2004. Like most people who have worked for the same organization for decades, he was due a pension. But unlike most retirees, this did not involve a trip to the local bank or welfare office. Instead he had to go to South America and carry out a brush contact once a year with a Line N officer from the Russian Foreign Intelligence service. This annual trip became his only operational act (one the FBI carefully monitored). On August 25, 2007, the FBI watched and videotaped a meeting in the same Peruvian park. He was seen meeting a man, later identified as a Russian embassy employee. They walked together and then sat on a bench. The Russian then placed a shopping bag into a plastic bag held by Lazaro. On his return, Lazaro spent eight thousand dollars on county and city taxes.

It is telling that after retiring from three decades as an illegal, Lazaro continued to live in America. This was a man who had been given the mission of spying on the country. And yet, even though he was no longer an active spy, he decided to stay in the country rather than return home. He was not, it is thought, unique in doing that. Other illegals were also allowed to stay in the United States after a quarter of a century's service during the Cold War, SVR officers would say. The only conclusion to draw from this is that they simply preferred life in America, especially as a retiree and especially if, like Lazaro, they had put down family roots. This was always the lurking fear at the heart of the illegals program when it came to Western countries. Lazaro's career was over and the era of the old Cold War family illegals was also slowly drawing to a close. Now a new breed of spies was on their way to the West. And they would look very different.

11

Enter Anna

THE YOUNG RUSSIAN was hard to miss. She was wearing a long, flowing white dress when Alex Chapman saw her at a dance rave in London in the summer of 2001. The dress, bought for her school graduation, reached down to her ankles but did not hide her curves. Alex plucked up courage and went over to talk to her. "I'm sorry, but you're the most gorgeous girl I've ever seen," the tousle-haired, easygoing English boy said. She did not brush him off. Her name, she said, was Ana Kuschenko. She was nineteen and from Russia. Alex was twenty-one. He loved music and was working at a recording studio after leaving boarding school at sixteen. He could not believe his luck. The pair talked until seven the next morning. He was smitten. She would later claim it was love at first sight. This would have been a disappointment to Marcus Read, who had met her the previous summer in Africa. She had then called from Russia saying she wanted to see him and asking him to sign some papers to invite her over. He recalled they had gone straight to bed when she had arrived in the United Kingdom but now, six weeks later, she was hooking up with Alex Chapman.

Ana was studying economics in Moscow. She had grown up in Volgograd—teachers remembered her as shy and well-behaved, largely looked after by her grandmother as her parents were abroad. She always wanted to be a businesswoman, they would later say. She was from the first generation of Russians for whom the Cold War was history. They could enjoy moving freely between Russia and the West. Ana rang Alex the day after the rave and they met up. She said she loved his accent. And the hair—a bit like Liam Gallagher from the rock band Oasis. The next day she flew back to Russia and there were tears. But Alex decided he would pursue her to Moscow. There were trips back and forth. On a British Airways flight to Moscow in January 2002 he would say they joined the mile-high club in the bathroom. He was totally infatuated and asked her to marry him. They tied the knot at a registry office in March 2002. He rented a dinner suit that sat loose on his skinny frame. She wore the same white dress she had worn the night they met, just over half a year earlier. Ana later denied that there were any "secret motives" for her marriage and said that it had always been love at first sight. But marrying a Briton or an American was quite literally a passport to the West for a young Russian woman. Now Ana had hers. Good-bye, Ana Kuschenko. Hello, Anna Chapman.

Anna Chapman's story has been the stuff of dreams for tabloid newspapers in London and New York, the two cities where she lived, with the pictures to go with it. But that has done something to obscure her significance. For decades the illegal program had been on autopilot—doing what it had always done in sending out deep-cover illegals to pretend to be someone else from another country and slowly embed themselves among their targets. But Chapman would be part of a new generation of undercover Russian intelligence officers—what the FBI called True Name Illegals. And whereas the other "family illegals" would increasingly represent the past when it came to Russian illegal activity in the West,

Anna Chapman was the future. One of the reasons was the way the world changed in September 2001.

A few weeks after Anna Chapman had come to Britain and met Alex, Al Qaeda struck. The day after the 9/11 attacks, CIA counterterrorism officials briefed President George W. Bush. They said the Russians could now be allies against Al Qaeda. When CIA director George Tenet went to talk to the leadership of Russia House the next day about how that might work, officials there were "dumbfounded," two of those involved, John Sipher and Steve Hall, later wrote. "To those of us who had worked on Russian issues for years, we knew that there was no way the Russians would be real allies." Russia House's deep skepticism of the possibilities of liaison remained. Putin thought 9/11 meant that he would now have support for what he saw as his battle against terrorism in Chechnya. He would be disappointed. One of the recurrent themes of their relations was how little the two sides understood each other. Putin never appreciated how 9/11 changed America and how it would react to that shock. He thought it would make the countries natural allies. Instead, under President Bush, it led to an America determined to assert its power.

The attacks meant the CIA and FBI (as well as MI5 and MI6) would pivot hard toward terrorism as their overriding priority. Everything now was about stopping the next attack. The officers who dealt with terrorism were now calling the shots. Those dealing with Russia—already seen as dinosaurs in the 1990s—were even farther down the pecking order. Old-school espionage seemed a thing of the past. But while the focus of the West and its intelligence agencies shifted over the years, darting around to meet the latest threat and now fixated on terrorism, their Russian counterparts, in contrast, kept their gaze firmly fixed on their old adversaries throughout.

The shift of focus toward counterterrorism by the FBI, CIA, and others might sound like good news for the Russian illegal program, drawing away attention. But it was not quite so simple. The attacks had one important consequence for Directorate S. The 9/11 hijackers had gotten into the United States all too easily and their presence had revealed the laxness of America's border security, visa, and identification systems. To stop further attacks, controls were going to be tightened up. There would be new systems and databases that could be cross-checked. In the past documentation was often a local affair. "It was literally paper copies with little old ladies flipping index cards, pulling them out and making photocopies," explains one counterintelligence officer. Now it was all going to be modernized to make sure the "dots" could be connected between who someone was and what the authorities knew about them.

This new world posed a significant challenge for Directorate S. It may have been aimed at terrorists, but the new checks had consequences for illegals. Donald Heathfield, Cynthia Murphy, and the other illegals all had their legends built before September 11, 2001. Many of the loopholes they had taken advantage of—both in the United States and Canada—to steal people's identities were being closed. Electronic databases were linking up different aspects of people's identities. If someone died a flag would now be put on his or her birth certificate. That meant stealing the identities of deceased babies to create dead doubles was going to be much more challenging. Traveling was also going to become more laborious as identity checks were increased and passenger data shared in advance so it could be cross-checked against databases of suspects.

IN THE EARLY to mid-2000s, another threat emerged not just to Russian illegals but to all spies—biometrics. This offered the chance to verify someone's identity using their unique characteristics. Databases of fingerprints and iris scans at borders make it almost impossible to enter as one person and leave as another or visit using multiple identities. The days of using "thin" cover by picking up one of the passports in your safe and then using it to go into a country and quickly meet an agent and then leave was passing. Even as far back as the early 1990s, Directorate S told agents to look for people who might have access to Western DNA databases. If people who had this access could be recruited as agents, it might offer the SVR the chance to tamper with the entries, insert false ones, or use the profiles of others. Now the checks were being implemented internationally.

On top of biometrics, the digital world emerging in the mid-2000s added another level of complexity. In the past, spies could create a legend pretty easily with some false documents. But now they also needed a digital backstory—Facebook accounts and a trail of online activity. If someone has no social media or digital footprint, that in itself looks suspicious and might look to a security service like a potential flag for being a spy. MI6 began to understand this in the mid-2000s. They ran tests to see how long a spy working on traditional cover could stand up against a suspicious border officer who was armed with nothing more than Google. The answer was about a minute.

Different intelligence services would be caught out by this new world. The CIA had a group of its officers involved in kidnapping a cleric in Milan exposed after they were tracked by Italian authorities who simply cross-referenced recently purchased phones with airline and hotel bookings to identify a rendition team. Meanwhile, a Mossad hit team in Dubai operating undercover was identified using closed-circuit TV and passports a few years later. The spying business was changing fast as—like other fields—the internet disrupted the old model. This new world of databases, biometrics, enhanced checks, and digital trails meant it was going to be much harder to create a sustainable backstory for illegals.

The FBI did not see any new deep-cover "family illegals" arrive in the United States after 2003. They believe the reason was that the new environment was making it harder to build a sustainable false identity. But now there were other alternatives and new methods for Directorate S as it began a renewed drive to spy on the West under Vladimir Putin. It could take advantage of the new openness of the West to Russians since the end of the Cold War. It had become far easier for students and businesspeople to come over. So one answer was to use people who no longer posed as someone else. Instead they would use their true names and real documents. These were called True Name or Special Agent Illegals by the FBI. Enter Anna. "She wasn't hiding what her name was—she wasn't hiding that she was Russian. What she was hiding was that she was working for the Russian Intelligence Service," says Maria Ricci. How did she get spotted by the SVR? That is not hard to work out. Alex Chapman would realize that the girl he had met had an interesting family.

The couple's honeymoon in the summer of 2002 was a long and unusual one. First two weeks in Egypt and then six in Zimbabwe. The latter was because Anna's father was based there. Anna's mother had decided she liked Alex, but when her father met him for the first time in Zimbabwe he was suspicious. Alex found him intimidating with his piercing eyes. The father relaxed a little when they went camping. He wanted to know what Alex would do to provide for his daughter. Her father was officially a high-ranking Russian diplomat, but he drove around in a blacked-out four-by-four with one car in front and another behind, more security than other diplomats. Only later would Anna eventually reveal the truth—Alex's father-in-law was a high-ranking KGB and now SVR officer. She clearly idolized him. "Her father controlled everything in her life," Alex Chapman later said. "I felt like she would have done anything for her dad." The honeymoon was action packed—skydiving, boats, a plane ride over Victoria Falls, and a safari. They returned to London and rented an apartment. Strangely, Alex then became a director of a company that transferred money from the United Kingdom to Zimbabwe.

For the next two years, Anna went back and forth to Moscow to finish her master's degree, graduating with first-class honors in 2004. She was clearly smart. But the marriage lasted only two and a half years. Anna was the one to break it up. Something changed. Already by late 2003, she was attending dinner dances at the Ritz—dancing, one person would remember, with a wealthy Texas businessman in his sixties. "There were lots of women there, but she stood out. She was wearing an incredible red dress," a lawyer later told a newspaper. The lawyer managed to get her phone number and took her out to dinner at upscale restaurants. "Anna had seductive charm," he said. "She was quite simply gorgeous. I didn't fall in love with her, but my God she had the magic."

Alex watched helplessly as his marriage fell apart. "There was such a dramatic change in the way she thought and the way she went about things. I felt I hardly knew her anymore," Chapman said later. "It was like someone having a midlife crisis, but in their 20s. She would arrange to go out but when I said I would join her she told me not to bother because they would all be speaking Russian. She was adamant I wasn't to meet them. She had never been materialistic during the years we were together, but in 2005 and 2006 after she started having these meetings with people she referred to as 'Russian friends,' she was transformed into someone with access to a lot of money, boasting about all the influential people she was meeting."

These were the years when the British capital came to be known as Londongrad. From the 1990s, a wealthy Russian community had embedded itself in London life. Some oligarchs and businessmen had left Russia for good, but many just used London for a brief escape. It was a short plane ride; the visa regime was permissive and there were plenty of places for a wife to shop while a business deal was done. And there was the fact that the city of London was the number one destination of choice for dirty money around the world. There was plenty of it flowing out of Russia. Hundreds of billions of pounds of criminal money is laundered through British banks each year, aided by a regime of "light touch regulation." Some estimates put the amount of Russian money coming to the United Kingdom as 100 billion pounds over the last twenty years. Money bought influence and the existence of a large Russian community provided new opportunities for Russian intelligence to operate. They also had little to worry about from the authorities in those years. One MI6 officer was astonished when he discovered that the phone line of the SVR residence in London was no longer monitored 24/7. And after 9/11, Russia dipped even lower down the priority list. The resources directed against counterespionage in MI5 plummeted to just 11 percent of its budget in 2003–2004, its head acknowledging "significant risks" were being taken as a result. One man who could see what was happening was Alexander Litvinenko. After he had arrived in the United Kingdom, he was interviewed by MI5 and MI6. One thing he warned them was that he could see the Russian organized crime world he had tracked as an FSB officer in the 1990s was now exporting its work into London.

Anna Chapman was part of a new type of espionage that took advantage of the West's and especially the United Kingdom's openness to those coming over in a way that would not have been possible in the Cold War. Russia was once again building up its espionage capability—both new and old. Soon after Putin took power, spy fever at home was twinned with more aggressive spying abroad, particularly in Europe and the United States. The SVR was given a direct order from Putin to step up its work. One report said the number of SVR officers operating out of London rose from a lonely single individual in 1991 to thirty-three by 2002. By 2005, MI5 was warning that there were at least thirty Russian spies under diplomatic cover. Soon the number would exceed that seen at the end of the Cold War. Some of those spies (who make up more than a third of embassy staff) were carrying out traditional intelligence work, spotting and slowly cultivating agents with access to secrets (always avoiding phone calls, and perhaps with a traveling illegal meeting them abroad in Cyprus or Austria if they are recruited). But there were changes as well—like spying on exiles who had set up home in London. Watching these individuals became the new priority for Russian spies abroad. The Kremlin was convinced that émigrés—like Berezovsky—were plotting campaigns of subversion within Russia and they needed to be monitored—and perhaps dealt with. In London, arch-critics of Putin mixed with those who supported the regime or who had more ambiguous relationships—creating both the desire and opportunities for spies to watch what was happening. And Moscow's spies found it easy to operate. Sometimes one would be under surveillance from A4—MI5's watchers—but it would later become clear the Russians had known it and were actually watching the watchers, observing their tactics and procedures so that when the need arose they could go dark and avoid surveillance. "It was our C team against their A team," is how one senior British counterespionage official remembers it. And now London was full of Russian businessmen and young Russian women. The days of being able to follow a few Russian diplomats around and feel like you had a handle on potential espionage activities were long gone. How did you know who the spies were when there were so many Russians among you?

• • •

AFTER SPLITTING WITH Alex, money was initially tight for Anna. She worked at a branch of Barclays Bank and moved into a small apartment with a young woman from Belarus. "My dream was to go to London and find a rich husband," that friend later told a journalist. "Anna taught me how to be hard-nosed about these things. She would go and talk to a man if she thought he was useful." The friend remembers a succession of men surrounding Anna. "Whenever she met a man who interested her, she would always get his phone number, always, it was amazing. I took her to church once, and at the end she even got the number of the priest!" Anna started a series of jobs that placed her in the fast set of London business and life, like one for a company selling private planes to companies and individuals in Russia. She threw herself into the new world of the international jet set who had made London their home and began attending film premieres and society parties.

In 2005, Nicholas Camilleri, who ran a hedge fun called Navigator, was introduced to Anna Chapman by a London socialite at the restaurant Cipriani. Anna was looking for a new job and she became his assistant. "She was very beautiful," he recalls. "She got along well with people and people liked her." She would accompany him as part of his entourage as they hit the party scene for the next year. He took her to places like Annabel's—long the hangout for minor royalty, Euro-trash, and assorted hangers-on. She got herself close to a number of well-connected international socialites and a friend would later claim she had been specifically attending Boujis nightclub to meet Princes William and Harry, who partied there, although Camilleri is skeptical of such a possibility. Next came a relationship with a Frenchman and the chance to move into his flat just off the King's Road in Chelsea—the other side of the city, literally and metaphorically, from the small flat she used to share with Alex in Stoke Newington. By 2006, she was on the committee organizing the 235-pounds-a-ticket, white-tie, Russian-themed War and Peace Ball at the Dorchester. The patrons included a long list of European aristocrats. Attendees included the Russian ambassador as well as minor European royalty.

Camilleri remembers that one of his contacts in particular took an interest in the young Russian woman. Camilleri introduced Boris Berezovsky to Chapman one night. "She latched on to him," he recalls. "He took quite a fancy to her." Berezovsky would send his car across Mayfair to pick her up for lunch regularly. She would then come back to the Navigator office and repeat all the not very complimentary things the exiled oligarch had said about Putin. Berezovsky's political asylum in the United Kingdom rankled the Kremlin. Putin saw his presence as part of a conspiracy. Berezovsky would later boast that he was seeking regime change and had supported opposition groups in places like Ukraine (partly through his nongovernmental organizations) as a way of leveraging change into Russia. The Kremlin—perhaps unsurprisingly, since Berezovsky was allowed to operate in London—saw this all as something encouraged or supported by Western intelligence agencies.

So what led Chapman to be close to Putin's arch-critic? Wealth and power? Or orders from the center? The crucial question is when Anna Chapman became Special Agent Chapman. Camilleri doubts she was working as a spy during the time he knew her. But her ex-husband, Alex, would later claim that he suspected his wife had been "conditioned" by Russians in the United Kingdom at the time their marriage was disintegrating. He said she started to have "secretive meetings" with Russian friends and began to change. It is possible that this was the familiar story of a carefree girl becoming someone more interested in money and social climbing, who was then recruited by the Russian intelligence service. When FBI officials looked back over her trajectory, some think that she had already been recruited when she came to the United Kingdom and that her time there had been to legitimize herself and build her cover. In this way at least, it would be similar to the way Donald Heathfield had built his identity and cover in Canada and France before entering the United States. Others in the United Kingdom, though, believe she was recruited by the SVR in the years she was living in London, perhaps during her regular trips back to Moscow.

CHAPMAN AND OTHER new illegals were called "agents," since they were recruited by the SVR rather than having been selected and then put through the academy and trained up as full intelligence officers, like the "cadre" or family illegals. The selection, training, and cover of these agents was different. Normally they would be introduced to Directorate S by someone who had spotted them and told the SVR they might be both capable and useful. They would then be given some basic training and sent out into the field to see how they would manage. They would be given carefully selected operational tasks and if they showed they could cope, they would be given more training, a higher rank, and more advanced tasks. The advantage was that because their Russian heritage was not hidden they could travel back to Moscow easily for training and reporting. Using their true identity meant they could get around the problems other illegals faced with their false cover stories. Biometrics at the border were no longer a problem and when they did want to go back home there was no need for the complex routes via third countries and picking up of false passports to hide their trail that the other illegals were forced to deal with. Why did you need costly illegals when you had a small army of businessmen and young people who now easily traveled back and forth to the West? This was the significant shift within

Directorate S—the realization that the new flow of Russians into the West offered new opportunities.

One of the secrets within the KGB and SVR was that their legendary illegals were limited in what they could achieve. The "backstopping" (false backstory) of such officers was never going to be good enough to allow them to do what the Russians really wanted—directly infiltrate foreign governments and intelligence services. But still they were part of the mythology of the KGB and then SVR. Their presence allowed Russian spy chiefs to tell their president that they had people living hidden among the adversary, unseen and ready to follow orders. "That was of great comfort to them. It was almost in their DNA—they have to have these people throughout the world," says one senior Western counter-intelligence official, "but now when you flash forward into the twenty-first century that model was not really a successful model and so they had this other replacement model that was going to be much more successful and more importantly didn't have limitations that the cadre illegals had." Rather than spending years training a Russian to look like a Westerner, you could now just have a Russian open an art gallery in London's Mayfair or Manhattan's Upper West Side, which could attract all kinds of interesting people through its doors at receptions and whom you then might be able to report back on. Another difference was that the new special agents tended to operate alone rather than in couples. The old illegals had been sent as a pair since falling in love with a local risked blowing your cover. But an illegal who was single offered different possibilities.

Anna Chapman would often be portrayed in the media as a "honey trap"—sent to seduce. The notion of Russian honey traps is part of popular culture thanks to films and books like *Red Sparrow*, the fictional account of a Russian ballerina trained by the SVR to seduce. The use of honey traps is well established as a technique to target foreigners within Russia (and also some countries where the Russians feel comfortable operating). Despite all the warnings, many a middle-aged diplomat has lacked the self-awareness to appreciate that the attractive younger woman he has met at the hotel bar is not actually interested in him for the reason he hopes. The video he might be shown a few days later will make that all too clear. A Marine Guard at the US embassy was compromised by a Russian woman back in the 1980s, and more recently, in 2009, a British diplomat was filmed with two women and had to leave the country. Men have also been used to entrap. One Russian in the early Cold War was used to target a ballerina in Paris who was friendly with American military officers. Men are also used to target other men. Because of the way homosexuality is viewed within Russia, they often think that entrapping someone in a homosexual relationship will provide leverage. Visitors to Russia are warned of the risks from both sexes. One FBI official used to warn members of Congress by telling them to go take a hard look at themselves in the mirror before they went to Moscow and rate their own looks. "If you are a solid 7 but when you travel everyone who comes up to you is a 10, you should be concerned."

But how extensively does Russia send "sparrows" abroad? Some believe Moscow is more likely to work opportunistically, keeping an eye out for any Russian women abroad who manage to get into a relationship with someone interesting and then approaching them. But others believe there is an established program. "Red Sparrow is not that far from the truth," one senior US counterintelligence official claims. This was often built more around developing long-term relationships rather than one-night stands. A former member of Directorate S says that in the 1990s, the SVR was sending out women as what he called "sleeping agents" who would marry influential foreigners with the aim of establishing deep cover. They would be sent for the long term, living under their real names as "auxiliary agents" emigrating as part of the wave of Russians leaving the country in the 1990s to explore the new openings in the West. They were never fully trained-up illegal officers or "Special Agent illegals" who had been recruited as agents and they would not have false documentation. There would be only limited contact with the Center, normally through a cover address, and perhaps only once or twice a year to check up on them. They might be activated a few years down the road, depending on where they had placed themselves. In a separate case in the 2000s, the FSB carefully maneuvered a former escort, alongside a well-connected, wealthy British businessman in London. She had been carefully selected to appeal to him, not just due to her looks but also by creating a backstory that would elicit his sympathy because of her personal plight involving a child. The goal was to establish a close relationship that could then create what is called a "platform" for further FSB operations; the woman could leverage his contacts to "talent-spot" for Russian intelligence in the influential circles he moved in. In this case suspicions were aroused among some of those close to the businessman. The woman's background was uncovered and the plan foiled.

But Anna Chapman herself does not appear to have been a "honey trap" or sparrow. She was a trained illegal agent, the daughter of an SVR officer. The fact that many men found her attractive was simply an added bonus that offered her additional opportunities in her mission as one of the new breed of illegals. "She had a particular set of skills. And she was able to use them quite effectively. The purpose for a honey trap is a different purpose to what she was doing," says FBI agent Derek Pieper. At the time she was blazing a trail through London though, MI5 had no idea that Chapman was operating on behalf of Russian intelligence. After all, how did you know which of the thousands of young Russian women were the dangerous ones? When Boris Berezovsky was asked about Chapman after it was later revealed she was a spy, he denied being particularly close to her. "There were many Annas in London," he said. On one level that was true. There were many young Russian women in the city. But there was only one Anna Chapman.

IN LATE 2006, Anna told Alex she was going back to Russia for good. The plan seemed to be for her to set up an online real estate company. But then she returned to London and now all her talk was about America. This sounded a little odd to her former husband. When they had been together she had always sounded pretty anti-American, turning her nose up at Hollywood films. She soon started making on-and-off trips over to the United States.

At the same time as Anna was beginning to think about moving on from London in 2006, British intelligence was faced with the sudden realization that Russian espionage had not just continued but was far more dangerous than they had understood. They realized it because of one of the most extraordinarily aggressive acts in the post–Cold War era by the Kremlin—the murder of a former Russian intelligence officer in London using a radioactive poison.

12

The Spectre

ALEXANDER LITVINENKO WAS a spectral presence lying in a hospital bed in University College Hospital, London, in November 2006. Ghostly pale, he had been vomiting blood. His immune system was failing, his heartbeat irregular. When his wife, Marina, an elegant former dancer, had stroked his hair to comfort him, she stared at her hand in horror as clumps of his hair came away and now it had all been shaved off. His body was collapsing from within, but his mind was strong, and he had a story he wanted to tell. He knew he had been poisoned. But no one would believe him.

Litvinenko had fallen ill at home in North London on the night of November 1. It had been the anniversary of his family's arrival in the United Kingdom, so his wife had cooked his favorite chicken. But soon after going to bed he had rushed to the bathroom and vomited. Then again twenty minutes later. And again. There was foam and blood. In the early hours of the morning he told Marina he thought he had been poisoned. But when an ambulance came the next day, the medics said he was just suffering a bacterial infection and needed to drink more water. Things got worse. When another ambulance was called the next day, he was finally taken to Barnet Hospital, in the North London suburbs.

When he explained he was a former KGB officer who had been poisoned, everyone thought he was crazy and they did not take much notice. For two weeks he lay there trying to convince people of his story, a story that would reveal how far Russian intelligence was willing to go under its new leader.

Two police officers were told to go see him on November 17. Their briefing was that there was a man making some wild claims. But it was probably nothing. On their way they were told his condition had deteriorated and that he had been transferred to University College Hospital, a modern tower right in the center of London. Just after midnight, a detective inspector from the Metropolitan Police's homicide team sat by Litvinenko's bed on the sixteenth floor. The medical team said there was no evidence of poisoning and gave the police the impression they were just getting in the way. His situation was deteriorating and no one could understand why. Can you guarantee he will still be alive tomorrow for us to talk to him, the police officer asked them? No, was the reply. Then the interviews would begin right away, the police insisted. The two officers sat in the room while an old-fashioned tape machine recorded the conversation. There was a uniformed officer at the door but no armed guard. "I have name Alesksandr Litvinenko. I am former KGB, FSB officer," Litvinenko explained in faltering English. That night was one the two policemen would never forget as they sat for hours listening to an extraordinary tale, like something out of a thriller.

Litvinenko began to tell his life story. The military. The KGB. Then the FSB. Then Putin. As he spoke, he was still alert and could occasionally walk around the room, although clearly in pain. At one point he stared out of the window, admiring the view over London that the hospital room offered. He understood in a way that almost no one else did that the city in front of him was one that teemed with Russians and with spies. Litvinenko had days to think about the events leading up to his illness and now he wanted to share every detail with the police. Soon it was past two in the morning. Wasn't he tired? You might be tired, but I'm happy to carry on, he told one of the police officers. Finally, at quarter to three they finished. The police officers were so intrigued by what they heard that at four in the morning they were standing outside a sushi restaurant in Piccadilly, one of the places Litvinenko had visited and which they thought could be the site of a poisoning. It was closed but they just wanted to see it themselves. They told their superiors at Scotland Yard they had to go back to the hospital and hear more.

At the end of one tape, the police officer asks a familiar question: "Can you think of anybody else who may wish to do this sort of harm to you?"

"I have no doubt whatsoever that this was done by the Russian Secret Services. . . . I know the order about such a killing of a citizen of another country on its territory, especially if it had something to do with Great Britain, could have been given only by one person."

"Would you tell us who that person is?" asked the police officer.

"That person is the president of the Russian Federation—Vladimir Putin."

When it comes to motive for why the Russian state—and even Vladimir Putin himself—wanted to kill Alexander Litvinenko, the challenge is not discovering a motive but working out which of the many possible motives is the most likely one. Alexander Litvinenko was a man who had personally confronted Putin about corruption and engaged in a bitter and personal battle with him afterward. He was a man investigating the ties of some of Putin's closest allies to organized crime. He was a man who publicly accused the Russian security services of murdering its own citizens for political gain and who had been working closely with the

Kremlin's nemesis in exile. Much of this could be found out by those who would look into the case. But there was also another, hidden aspect to his battle with the Kremlin that would remain largely unseen.

When he first fell ill, Litvinenko had asked Marina to bring two phones to the hospital. On one phone he made a call to another former FSB officer, named Andrei Lugovoi. Litvinenko explained he would not be able to go to Spain and meet Lugovoi there in a few weeks, as they had planned. He used the other phone to contact a man called Martin. Litvinenko did not understand that even though he was using two separate phones, those conversations were linked together in a way that would explain his fate.

At the next hospital interview, Litvinenko's condition was clearly worse. The pair of officers were now willing to believe that this was some kind of mysterious poisoning. They asked Litvinenko whom he met the day before he fell ill. Litvinenko recalled a meeting at 4 p.m. on October 31 in Waterstones bookstore on Piccadilly. But he would not provide the man's name while the tape was recording. "It could be absolutely vital you tell us who that person is," the policeman said. "You can call him and he will tell you," Litvinenko responded. The tape was stopped. It was over two hours before it was turned on again. In that time, the man whom Litvinenko mentioned had come and gone from the hospital room. He had embraced Litvinenko when he had arrived. The pair clearly knew each other well.

The police interview resumed with the officer asking if the person who had just been in the hospital was the man whom Litvinenko had met at Waterstones. Litvinenko said that he was. "I don't want to ask you what you talked about with that person," the police officer said. He understood this case had moved on to even more difficult terrain. The man had been "Martin" and he was from MI6.

There were many unprecedented, extraordinary aspects to this case for the police, but one was that they had in front of them what they would call a "living murder victim." A man who was dying but whom they could interview. As with any murder inquiry the key is to find the point of contact between a victim and the weapon that killed him. But rather than having to reconstruct his movements up to the point he fell ill, they could ask him to recount every minute. The fact he was a former FSB officer meant he was trained to be aware of his surroundings and of unusual behavior, so police officers found him a far better witness than most people would have been. Despite the huge pain he was clearly suffering, Litvinenko was diligent and dignified as he spoke, but still his condition was visibly deteriorating as the interviews went on. Some wondered if he was suffering from thallium poisoning, but the symptoms just did not seem right. The medical team could find no explanation. The case was close to being abandoned as no one could find evidence of deliberate poisoning.

The other tool in a murder inquiry is a postmortem. The senior officer in charge ordered what he later described as a "living postmortem"—a battery of tests and examinations from every possible expert. But they were all coming up blank. A Geiger counter had shown no signs of radiation. A urine sample was sent to the Atomic Weapons Establishment (AWE), which builds the United Kingdom's nuclear weapons. The lead police investigator assembled all the scientific experts. He held up two pictures: before the poisoning there was a man looking healthy and fit, and after he was pale and emaciated. "How did he go from that to this?" asked the officer. They went through all the possible options. Finally one person from AWE chirped up that they had found a tiny spike of polonium in a urine sample. You mean plutonium, asked the police officer? No, polonium, said the scientist. A liter of urine was needed to test properly, the scientist explained. "You are taking the piss," said the policeman bluntly, only later realizing the irony of his words. Litvinenko had stopped eating and drinking a week earlier and was producing no urine. Fortunately, a sample had been kept from earlier and was sent off.

What would be the last police interview came on the evening of November 20. Litvinenko was moved into intensive care that day. Wires and monitors were all around the now-hairless Russian lying there in a green hospital gown. As the interview was coming to an end, Litvinenko was asked if there was anything else he wanted to say. He had just become a British citizen a month earlier, on October 13. The ceremony was a proud moment. The same day there had been a memorial meeting for the crusading Russian journalist Anna Politkovskaya, who had been shot dead in Russia. "I just received my citizenship, now they will not be able to touch me," Litvinenko had told a friend that day. He thought British citizenship would protect him. He was wrong. Now on his deathbed he recounted to the police that he had taken his son to the Tower of London. "I showed him the British crown, and I told him, 'Remember for the rest of your life this country saved us and do everything whatever you might be able to do in order to defend this country.' " The police officers found themselves increasingly attached to this unusual Russian fighting for his life.

Litvinenko had one request for the police officers. He asked them to pursue the case wherever it took them—he said he knew there would be political interests that might get in the way, but he asked them to make sure the investigation that had started in that hospital room was completed. He would be right that politics would obstruct the search for the truth. But as the final tape finishes, the police officer makes the Russian a promise. "I will do absolutely everything within my power to ensure this case is properly investigated." The police would not let him down.

Alexander Litvinenko was beginning to surrender to the radiation that was cooking his body from the inside out. Early on the twenty-second of November he suffered two heart attacks, from which he was resuscitated. He kept fighting. On the afternoon of November 23, a call came into the police from AWE. The presence of polonium was confirmed. A few hours later, Litvinenko suffered a third cardiac arrest. This time his body gave up the ghost.

A statement that he had agreed to just before he died was read on the steps of the hospital the next day by friends. Journalists had gathered after reports had begun to emerge that there was a former Russian spy claiming he had been poisoned. "I can distinctly hear the beating of wings of the angel of death," the statement read, before going on to make clear who Litvinenko held responsible. "You may succeed in silencing one man but the howl of protest from around the world will reverberate, Mr. Putin, in your ears for the rest of your life. May God forgive you for what you have done, not only to me but to beloved Russia and its people."

There was a deeply personal element to the antagonism with Putin. It went back to when Litvinenko had gone to Putin's office to ask the then head of the FSB to clean house. In London Litvinenko had become increasingly vocal in his attacks on Putin under Berezovsky's patronage. In 2001, Litvinenko had coauthored a book called *Blowing Up Russia*, which pushed the theory that elements in the FSB itself were behind the bombing of apartment buildings in Russia in 1999. These bombings, the theory claimed, were designed to provide a pretext for a second intervention in Chechnya to cement Putin's rise to power. Berezovsky in London helped publicize the book. Berezovsky's and Litvinenko's activities infuriated Moscow. Putin and the FSB saw the former officer as a traitor. They even used pictures of Litvinenko for target practice at a special forces training center.

Litvinenko had seemed determined to aggravate Russia's leader. In July 2006, he published an article on a website that could not have been more personal in its criticism. He pointed to a picture of Putin kissing a boy on the stomach at a public event and accused the Russian leader of being a pedophile without offering any further evidence. All of this antagonism was why Litvinenko was so sure that Putin was personally to blame for his poisoning. But what would be the verdict of the British state?

Polonium was the best and worst murder weapon. The best since it was so hard to detect and had not been seen before. The worst because once it had been discovered it left a trail that could be followed. It was incredibly persistent and almost impossible to remove every speck. So as they searched for the point of contact between weapon and victim, the police were able to follow this trail that would eventually stretch from sushi bars to hotels to Emirates Stadium, home of the Arsenal football club. London watched transfixed as more and more sites were closed off and white-suited forensic specialists began their searches for radioactive particles. There was fear over the long-term effects for those ordinary members of the public who had come into contact with it. One place soon became the center of the inquiry.

The day before he fell ill, Litvinenko had drunk tea at the Millennium Hotel. It sits on one side of Grosvenor Square, just across from what was then the site of the US embassy with its proud eagle overlooking wealthy Mayfair and its busy CIA station buried deep inside. The hotel catered to wealthy tourists and businessmen keen to be near the best shops. The hotel's wood-paneled Pine Bar, just next to reception, was the kind of place where a few Russians would draw little attention. And while there were CCTV cameras elsewhere in the hotel, there were none inside the bar.

"There's still some tea left here," Andrei Lugovoi had said, indicating the pot on the small table that fateful afternoon. Litvinenko had taken the bus and tube up from home earlier and then had lunch with a contact at an Itsu sushi restaurant before arriving. The Pine Bar was crowded and he was there to meet Lugovoi to discuss some private security work they were undertaking together. They sat at a table in the corner. Lugovoi had been there for a while already—racking up a seventy-pound bill covering cigars and cocktails. He knew Litvinenko was hard up for money and the prices in the hotel bar were so ridiculous he must have assumed his guest would not want to risk ordering his own drinks and having to pay. And so Litvinenko poured out half a cup of green tea. He did not see anyone else drink from the pot. The tea was already cold and did not have the sugar he liked, so he had only a few sips. It was enough to seal his fate. But if he had had more, then the long, drawn-out death he would face might have been much quicker. And that would have left no time for the police interviews or the unusual tests. His death would simply have been chalked up as unexplained. As the police zeroed in on this encounter, they hunted for evidence. The detectives were told the six hundred cups and saucers and hundred teapots the hotel used had gone through forty-two washes since the visit. But three hours later, the tests came back with what an investigator described as a "nuclear teapot." They had found the murder weapon.

They were able to correlate the polonium trail with two Russians—Lugovoi, and also Dmitry Kovtun, who had been at that meeting. The airplanes they had flown in would show signs of contamination. At one point the team found so much nuclear material in the bathroom at the Sheraton Park Lane hotel that Lugovoi had used that the detectives literally ran away.

Lugovoi had joined the KGB's Ninth Directorate, which provided security for senior officials. In 1996, he left to work as head of security for a TV station run by Berezovsky. In 2001, after Berezovsky fled, Lugovoi was arrested for trying to help one of Berezovsky's allies, Nikolai Glushkov. Lugovoi was said to have served a fifteen-month prison sentence. One possibility is that this was when he might have been approached by the FSB to work for them and infiltrate Berezovsky's circle (perhaps in return for a shorter sentence). Others wonder if the prison sentence itself was a ruse to bolster his credentials. Whatever the case, his criminal past did little to hinder his new career on release, running a private security firm in Moscow. In 2004, Lugovoi had gotten in contact with Litvinenko. Litvinenko trusted Lugovoi, introducing him to his wife at Berezovsky's flashy sixtieth birthday party at Blenheim Palace.

The relationship seemed a potential godsend. Litvinenko's relationship with Berezovsky was on the decline. The money from his paymaster was drying up and Litvinenko was worried and casting around for new ways of making use of his knowledge. He had moved into commercial due diligence work. Companies considering going into business in Russia or doing joint deals wanted to know who their partners might be, and an array of British firms offered their help.

Lugovoi—a man who had good contacts in Moscow—was potentially a useful source of information and Litvinenko suggested they start working together. Lugovoi would later introduce Litvinenko to Kovtun. Kovtun had known Lugovoi since they were children, as their fathers had served together in the army. Kovtun had joined the army and been posted to East Germany, where he met his first wife. "He had all sorts of dreams and plans, none of which he realized, however," she later said. This included wanting to star in pornographic films. He also drank too much, she said. He had later become involved in technical surveillance—bugging—a useful skill set in the murky world of "due diligence"—checking out your business contacts.

Litvinenko had been asked by a private security company in the summer of 2006 to produce a report on a former KGB man who had risen with Putin from his St. Petersburg days and whom their client was considering doing a deal with. The first report Litvinenko handed over was only a third of a page long and he was told it was not good enough. The main author had been Lugovoi. Litvinenko turned to Yuri Shvets, a former KGB officer now living in America, who sent a more impressive eight-page report back. Litvinenko showed it to Lugovoi as an example of what was wanted. When Shvets found out, he was unhappy. "Do you understand what you are doing, because it may be dangerous for us both," he told Litvinenko. Who might it be shown to? And with what risks? The report led to the collapse of the business deal and losses for Putin's ally of perhaps ten to fifteen million dollars. Here was another motive for why someone might want Litvinenko gone—he had cost powerful people a lot of money. But peel away the surface and there were still more, hidden layers.

As the money from Berezovsky dried up, Litvinenko had turned to MI6 to work as a consultant. From around 2003 they put him in touch with European security services. He was paid two thousand pounds a month and MI6 gave him a passport and a cover name—Edwin Carter. The Russian mafia had put down deep roots in Spain and this proved the most fertile territory for Litvinenko as he started to travel there from late 2004. The Spanish judge leading the investigation would later say that Litvinenko's "thesis"—that the intelligence agencies controlled organized crime in Russia—had proved accurate as the FSB was "absorbing" the mafia and eliminating noncompliant bosses. Just before he was poisoned, Litvinenko suggested Lugovoi could meet some people in Spain whom he was working with there. This was why he had phoned Lugovoi from the hospital before he had fallen ill to say that the trip was off.

And there was another, final, crucial layer to the relationship with Lugovoi that was swept under the carpet. Litvinenko was keen to prove his worth to MI6. He had not been an MI6 agent when he was in Russia but had been debriefed when he left and then became an "access agent"—someone who does not provide secrets themselves but helps provide access to people who might. A major in the Russian tax police claimed Litvinenko in 2002 had introduced him to an MI6 officer who started paying him two thousand euros a month for "consulting services." Meetings with MI6 took place over whisky in third countries like Turkey and Finland with Litvinenko sometimes attending, he claimed, as he was provided with a mobile phone with a special SIM card. Litvinenko was working for MI6 but those who know the details say he was not always easy to control and was always keen to do his own thing and prove his worth. Senior MI6 officers from the time speak with regret and concede they were not aware enough of what those beneath them had been doing with Litvinenko and the risks the activity entailed.

Litvinenko was trying to provide access to Andrei Lugovoi for Britain's spies. Lugovoi himself would later talk about being approached by MI6. These claims would be initially dismissed as attempts to muddy the water but one British source with knowledge of events confirms that there was an attempt to recruit Lugovoi. Lugovoi's meetings with Litvinenko to discuss business consultancy had taken place through late 2005 into spring 2006. "It began to dawn on me, that all was not what it seemed," Lugovoi later said, claiming he had been overpaid for his minor consulting work through an offshore company in Cyprus. "I was alarmed, because it was public domain information, which could be easily found on the internet. It became clear, that the purpose of the remuneration was to involve me gradually into cooperation." He said the British were interested in his contacts with security services and what the FSB was up to in the United Kingdom. They were even after information on people who might be close to Putin. "They started to try and recruit me openly as an agent for British Intelligence." Lugovoi claimed he was given a special phone by MI6 that he was to use when calling from Moscow.

Lugovoi would claim he was no admirer of Putin but says he was taught to defend his motherland and not to betray it. "When the British agents started to approach me, one of the first things that I did was to inform the FSB so that they wouldn't accuse me of being a traitor or a spy," he later said. And so now, the FSB knew what Litvinenko was doing for the British and that he was approaching people to try to recruit them. Litvinenko and MI6 thought they were cultivating Lugovoi with a view to recruiting him as an agent. Instead, they would be outplayed by the Russians. Litvinenko's target would become his killer. MI6 had failed to appreciate that the Kremlin would contemplate murder in London. Now its headquarters in Vauxhall Cross were in deep shock. "It was a kick in the guts," one British intelligence source closely involved in the case told me. "We never thought they would do it." The doubling back of Lugovoi was all part of the game the Russians had been playing for decades, but that game had normally been played on Russian territory and for lower stakes. Now they were playing it on the streets of Britain and exacting the ultimate punishment for those they saw as traitors even when they were abroad. Russian intelligence was changing and the old Cold War rules of the spy game were no longer in play.

Just because the murder weapon was so advanced, that did not mean the men using it were competent. Both Lugovoi and Kovtun have consistently denied any role in the killing but police would eventually conclude that there had been multiple attempts before the Pine Bar of the Millennium Hotel. A week before, Lugovoi had dropped a container holding polonium in his hotel bathroom and had used hotel towels to clean it up. And in one of the most extraordinary pieces of evidence, a friend of Kovtun's who ran a restaurant in Hamburg said Kovtun had asked him if he knew a cook, because he had a "very expensive poison" and he needed a cook to "to put poison in Litvinenko's food or drink." Lugovoi and Kovtun, it seems, had no idea that the poison they were carrying was radioactive. Lugovoi's own family was with him at the Millennium Hotel and he had his son shake hands with Litvinenko after the poisoning.

It was always implausible that this was a private hit. If the Russian mafia wanted you dead, they shot you or staged an accident. They did not use a rare nuclear isotope, like polonium, that could only come from a state's nuclear program. Britain should not have been blind to the new aggression coming out of the Russian security services, either. The year Litvinenko was killed, a new law had been passed in Russia. It gave the FSB the right to kill terrorists and other "extremists" abroad. It came in the wake of the brutal killing of a group of Russian diplomats in Iraq. But treachery was also in their sights. A colonel in the FSB who was Litvinenko's superior told a reporter years later what he thought should happen to traitors. "For me, a traitor, you spit on them, grab them and shoot them. Or hang them and piss on their grave."

Russia had killed a British citizen using radioactive material, leaving tiny particles scattered across London. The West was obsessed in these years about Al Qaeda and terrorists getting hold of "weapons of mass destruction"—even going to war with Iraq under the pretext of preventing that possibility. And yet the first use of nuclear material as a weapon against civilians was carried out not by jihadists but by Russian spies. And the response was . . . feeble. More than half a year after the murder, four Russian diplomats were expelled. And this was only in response to the failure to extradite Lugovoi rather than the result of any accusation of Russian state involvement in the murder itself. It seemed as if the British state did not want to confront the truth of what had happened. And so it did its best over the coming years to bury it.

The Cold War was past and the attitude in the Foreign Office was that the two countries needed to get back to a normal footing and put all this spy stuff behind them. By September 2011, Prime Minister David Cameron was visiting Moscow with the aim of boosting business ties. The following year Putin was in London sitting awkwardly alongside Cameron as they watched a judo game at the Olympics.

The British state did its best for years to block an inquiry into Litvinenko's death. The revelations might damage national security, it was said, including relations with Russia. It was only the tenacity of Litvinenko's widow, Marina, that kept the fight going, year after year, even after the money for the lawyers ran out. Eventually the government relented and allowed a public inquiry under an independent-minded judge, Sir Robert Owen. He was allowed to see some of the intelligence. His conclusions were powerful, making clear that Lugovoi and Kovtun were acting as part of an FSB operation. "Taking full account of all the evidence and analysis available to me, I find that the FSB operation to kill Mr. Litvinenko was probably approved by Mr. Patrushev [the head of the FSB] and also by President Putin," he concluded.

And so, a full decade on from the murder, this was at last the conclusion that many had been expecting. But what was the government's response to confirmation that Russia had carried out an act of nuclear assassination on the streets of London against a British citizen? Nothing. Given two opportunities, two different British governments had failed to take any significant action that might deter the Russian state from thinking it could murder on the streets of Britain with impunity. The feebleness of the response had meant that those in power in Moscow would not be deterred from trying to act again. There were multiple attempts on the life of Boris Berezovsky in London. One came in 2007—after Litvinenko's death—when MI5 and police followed a Chechen criminal arriving in London who Britain believed had links to Russian intelligence. This—one police officer said—was going to be the single most expensive "hit" ever ordered in Britain, with a price tag potentially in the millions. The assassin had brought his son as cover to make it look like he was a tourist, but the police were watching and arrested him before he could do anything. Amazingly, government officials were worried about the diplomatic fallout of making a public arrest. And so the assassin was quietly deported back to Russia. Soon after he returned, he was outside a restaurant in central Moscow when two armed men forced him into a car. He was never seen again. Failure has a price. There would be other suspicious deaths. In 2013, Boris Berezovsky would be found hanging by a scarf from a shower rail at a house in Britain. The coroner recorded an "open verdict," saying the evidence was contradictory, which left him unable to conclude whether the Russian had taken his own life or been unlawfully killed.

Why had the response to Litvinenko's murder been so weak? It's partly because every political leader—as in America—believed they could be the one to win over the Kremlin and put relations on a sounder footing. Then there were always the institutional voices saying it was time to move on from the past and put this spy stuff behind us and get back to business. But there were darker reasons as well. Russia had greater freedom of maneuver in London than in the past. It was not just that the city had become a playground for their spies. It was also the way that Russia now could wield a subtle influence in British life. The end of the Cold War brought an end to an ideological struggle. Now money—and greed—was the common denominator for many on both sides. In the West, the way in which Russia had opened up in the 1990s to businesses had drawn in a flood of Westerners on the make—all eager to cash in on what they saw as a gold rush. They were juicy intelligence targets for the FSB and SVR (and some of those individuals would rise to positions of political influence on their return). And once a form of oligarchical capitalism had taken hold in Russia, wealthy Russians then began to move their money to London, offering Moscow another way to exploit Western openness and greed. Money could be used to buy friends and protection. The year before the Litvinenko killing, a British intelligence official confided to me that one of the things that depressed him most was the way in which Russian money had "corrupted the professions" in London. By professions he meant those like law, accountancy, and banking. He did not mean the kind of thing we normally associate with corruption—direct bribes to break the law. It was a subtler type of corruption, he explained. The amount of money that Russian oligarchs and businesses could offer to lawyers, accountants, bankers, asset managers, and members of Britain's professional class meant they were willing to turn a blind eye to the origins of that wealth and not ask too many questions. It was an insidious form of corruption that was eating away at the core of public life. A whole class of people had emerged who provided professional services to wealthy foreigners and who in turn became wealthy themselves. They were influential people with influential friends, and they were now invested in making sure that the situation continued. In many ways this was just as dangerous and corrosive as the kind of bribery and intimidation that happened in "corrupt" countries.

Rich Russians had also learned how to launder their reputations through British society. Buying a football club, sponsoring an art gallery launch, organizing charity balls—all proved useful ways of winning friends and favors. Young, glamorous Russians like Anna Chapman were making their mark. The Kremlin was beginning to understand that influence offered a sort of power. It was not the "hard power" of the military, nor so-called soft power, when people are attracted by your values. It was a kind of dark power that worked unseen and played on greed and ambition. They could support friends covertly and make sure there was money and relationships at stake if anyone wanted to rock the boat with Moscow. In this way, the West would become complicit in the spread of Russian influence, opening new avenues for its spies to operate.

From around the time of Litvinenko's killing, British intelligence officers began to see signs that Russia was using money for influence further afield as well. There were reports that Moscow was using its energy industry to create dependencies in European countries. Lucrative contacts could be used to enrich people and put politicians in Moscow's pocket. No one was interested in the initial intelligence about Russia exploiting its new power at this time, they recall. This was all profoundly different from the Cold War days, when the two sides were largely cut off from each other. The Kremlin and its allies increasingly understood how the United Kingdom—and especially London—worked all too well and used greed and social status to embed itself in life in a way that in turn would make challenging Russia harder. This was the city that Alexander Litvinenko had looked out over from his hospital window and the reason why he had pleaded with the police officers not to give in to the pressure he knew they would face.

Why was Litvinenko killed? He was a former Russian intelligence officer working with the enemy. There was a lesson for Britain about what the Kremlin was willing to do and how far it would go to target those it considered traitors. The Russians were more ruthless and willing to go further than London imagined. But this was a lesson Britain and its spies failed to learn. The lesson Moscow learned meanwhile was that Britain was not willing to risk its own wealth to stop them. If you want a visual representation of what the influx of foreign wealth—much of it ill-gotten—has done to London, then take a drive in the evening around its most exclusive streets in places like Knightsbridge and Belgravia. The huge mansions—costing millions and bought in the name of shell companies registered by British lawyers in British Overseas Territories—lie dark and dormant. No one is living there. In these places, but also in its heart, London has become a ghost town.

13

Moscow Rules

In Moscow, Alexander Poteyev had come back to a Russia that was changing. His return from New York coincided with the emergence of Vladimir Putin in 2000 as a leader determined to lay to rest the chaotic days of Russia's past turbulent decade. And part of that meant dealing with traitors. Spy fever was gathering pace. For Poteyev, the path that he had embarked on was one of intense pressure. Every day could be your last. It would have been a nerve-shredding existence. And yet he operated as an agent in place for a decade. Working in Department 4, which dealt with the Americas, he had to evade the counterintelligence checks run by the directorate's Department 9 (much to the annoyance of the FSB, which would have preferred to know what was going on there). They were constantly looking to see if there was any sign that illegals operating abroad were under observation or had been turned. He knew all too well that one false move—or someone in Washington selling him out—would mean the end. There was a jittery atmosphere within Yasenevo's Directorate S, former officers remember. Everyone knew that a traitor could destroy their work in building up a new cadre of illegals around the world.

Poteyev had been recruited by the FBI. But running an agent

in Russia was not its responsibility. The streets of Manhattan were theirs, but Moscow was another story. That required the CIA. When an agent recruited inside the United States goes back to their home country, running that source should shift from the FBI to the CIA. But getting the transition of an agent from bureau to agency handling is not always easy. Often timing is a problem. It is no good telling us the week before someone goes back, CIA officials regularly complained. That was because it took time to build an operational plan on how to contact the agent securely and discuss it with him or her to make sure it was viable. They prefer to have a good six months, even a year, to be able to plan this transition. In the Poteyev case, a close working relationship between the two sides marked the case from beginning to end, leading to a successful handover. Each case is different, and a careful assessment needs to be made as to whether an agent is up to the task. It helps if the agents are spies themselves and understand the game. It takes the highest degree of skill to do so under the watching eyes of the KGB and then FSB. "It is like playing in Yankee Stadium," a CIA officer says. To succeed you need to do everything right the first time—operating under so-called Moscow Rules. One mistake and it is over. Expulsion for a CIA or MI6 officer. But a long prison sentence or perhaps death if you were the agent they were running.

For a Russian working for the CIA or MI6 in Moscow, the first challenge was trying to keep your identity secret. Living under constant surveillance plays with your mind. Western spies say they were living in a form of "reality TV" before the concept was invented. It was like the movie *The Truman Show,* in which your daily life was being watched and recorded and analyzed by an audience—how you ate your cereal and how you and your partner had sex were all part of the show. "So good to meet you in the flesh," one FSB officer said to a CIA officer. "I've seen so many videos of you."

• • •

THE DAILY SURVEILLANCE logs and eavesdropping reports for US diplomats would all be sent back to the American department of the FSB. And, for many years, the man sitting at the center of the web was Sasha Zhomov. The man who had once dangled the CIA as PROLOGUE took over as head of the American department in the late 1990s. At one point in his career he would work on Chechnya, but even then he never let go of watching the Americans. For the CIA, he was a figure who had faded in and out of sight like their own ghost. His job was to identify American spies but, more important, to use that knowledge to find the Russians who were being run as their agents. He and his colleagues studied the Americans in detail—looking to work out who the spies were and perhaps even find their own "hook" to target them. They pored over every file looking for clues as to who might be betraying Russia's secrets, determined to bring them down.

Within the American embassy, the atmosphere was claustrophobic. The array of tricks used by first the KGB and then the FSB ranged from bombarding embassies with signals to pick up conversations to spreading spy dust—invisible particles that could trace your movements. At one point in the 1980s, the CIA was so worried that its super-secure part of the embassy had been penetrated that it packed up everything, including furniture, and sent it back to the United States. CIA staff have to maintain their cover inside their own compound since some of the locally hired Russian staff were reporting to the FSB. Officers would carry out brush contacts within the compound itself to pass information or documents without being spotted. Getting out of the embassy to meet a contact was one of the biggest challenges. Everyone knew who the senior officers were, but a newer, young member of staff might be able to fool the other side. The FSB might put them under surveillance for a few weeks to check them out but then drop off, allowing them to try. Another trick was identity transfer, in which an intelligence officer put on a disguise that made them look like a "clean" member of embassy staff in order to get out without surveillance. For a while, officers would be smuggled out of the embassy in car trunks. Then the Russians realized this and would use sensors to detect heat. So something new would have to be tried. Techniques would cycle in and out of fashion as part of the cat-and-mouse game.

When you went out on the streets to meet an agent you had to be able to "go black" by carrying out a long surveillance detection route to get rid of watchers. The FSB had a scale of resources that dwarfed that of FBI and MI5. Sometimes the surveillance was designed to be obvious. It could even be helpful. In one case a CIA officer was sitting in a cafe near a train station waiting to catch a train back to Moscow. The surveillance officer actually came up to the American and reminded them that the train was departing in a few minutes because he did not want to miss it, either, and be stuck out of town. In another case an FSB surveillance team swooped in to prevent a CIA officer being subject to a mugging after their car tire was deliberately slashed. The relationships were a strange mixture of cooperative and collegial—each side making up nicknames for those they saw regularly but whom they rarely actually met. The episodes could occasionally be comic, like when a couple of CIA officers and their spouses decided to visit a museum out of town. The surveillance team followed them in only to find the museum was tiny and empty. The collected group of spies and their watchers found themselves being given an awkward guided tour together. There were unspoken rules, though. One was that if you messed with them, they would mess with you. And they had plenty of ways of doing that—from the simply unpleasant, like using your toilet and failing to flush, to the much darker, in which families were harassed.

Sometimes they would try to lull you into a false sense of security so that you would go out and meet someone and make a mistake, leading the way to your agent. Offense, though, has the advantage, those who have worked the Moscow streets say, because you have the initiative. All the training was so that when you were on the streets you were almost on autopilot—you were prepared for every possibility and would not have to overthink to react. It was alternately thrilling and terrifying, explains one person: terrifying for the moments leading up to any operational move but thrilling when you were in it as the adrenaline flowed. There was the sense that you were beating the opposition in their own backyard when they were holding all the cards. "There is no feeling like that—to defeat a really professional opposition," they say. When you were black, you were black. Then you knew you could meet someone. But it would have to be fast.

AN AGENT WOULD rarely be met face-to-face in Moscow. US sources will not comment on Poteyev, but the Russians believe he was able to carry out drops of information in the city. These may have been physical dead drops or they may have been electronic dead drops in which information is left or passed over an encrypted communications device (using these can be risky since they are a telltale sign of spying). These are useful for passing hard information but no substitute for a face-to-face meeting in which a case officer can talk directly to his or her agent, ask them questions, and also gain a feel for what kind of psychological state they are in and what is worrying them. But there was an advantage with Poteyev. He had to travel for his job. Over the next ten years Poteyev traveled to the Americas (mainly Latin America, including Mexico and Chile) twelve times on short business trips and a further nine times to countries neighboring Russia, according to an official Russian investigation. In the Americas, he would be officially visiting the Russian embassy in order to talk to their

Line N officers and check on their performance. In some cases, colleagues were said to later have recalled moments when they did not know where he was on these trips. This is thought to have been when he was secretly meeting the CIA face-to-face, under far less pressure than on the streets of Moscow.

CIA officers would head abroad under "non-official cover," posing as a businessman or some other professional to meet Poteyev. When they met, a CIA officer would often be accompanied by an FBI agent—including in Latin America. This was a carefully thought-through decision. The FBI, rather than the CIA, was the organization that was primarily using this intelligence and so would have the most direct understanding of some of the details and what questions to ask. There was also the fact that the FBI had cooperated so well in New York that it felt this was the right thing to do in order to keep the relationship ongoing. There was another calculation—it was thought important for the FBI to understand the risks Poteyev was facing and to see the challenges in trying to organize meetings. That meant seeing the whites of his eyes. It was hoped this would make it less likely they would make unrealistic demands and also remind people back in the bureau that one screwup—a messy covert entry into an illegal's house—and Poteyev's life could be in danger. The cooperation was tight. FBI officials were also embedded in the CIA's Russia House to receive messages coming back from Poteyev.

The intelligence passed at dead drops and at meetings was invaluable. Poteyev had direct access to the personal files of every illegal operating in North and South America. The illegals—people like Donald Heathfield and Cynthia Murphy—were not faceless agents to him but people he knew personally and whose daily lives—secret and public—he managed from afar. Every illegal was managed by someone back in Moscow who directed their work and kept a detailed file of their activities, cover, travel, legends, and intelligence product. Each three-hundred-page volume of the file took a year or two to fill with codes, telegrams, and reports. Poteyev had access to all this and could pass it on to the FBI. He was sending orders to the illegals and receiving details of what targets they were cultivating. In many cases, he personally would meet with them when they made one of their rare returns to Moscow. Bezrukov and Vavilov have said they knew him personally and met him a number of times. Bezrukov would later say Poteyev did not make a good impression and did not seem as professional as the others in the department, although that judgment likely involves a degree of hindsight. All his knowledge and insight, though, was being fed back to the FBI, allowing them to identify and monitor their targets.

Poteyev also picked up knowledge of other illegals operating in other countries around the world. This would also be fed back to the United States and lead to a complicated dilemma—if you knew of illegals operating in allied countries should you tell them? Those illegals would not be under watch like those in America were, which meant they could potentially be doing real damage to allied (and often American) interests. But what if telling them pointed the finger at your source? There is always this tension with the most sensitive intelligence sources—protecting the agent is vital, but intelligence is there to be used. In cases like the Enigma breakthrough in World War II, Britain pretended the material came from human agents to mask its true source. In modern cases, the opposite is often true. Intelligence is often portrayed as having come from intercepted communications when in fact it comes from a human agent. In practice, even if intelligence was carefully masked, if other countries acted on it, then it would get harder to avoid someone in Directorate S becoming suspicious that they might have a bad apple in their midst.

On January 25, 2003, Poteyev suffered a strange incident, the kind that makes a spy nervous. He lived in the Krylatsky Hills district, to the west of Moscow's center, an area of green parks. The drive to work at Yasenevo would have been just under an hour if the traffic was not too heavy. Just before 8 a.m. that day, his wife opened the door of their apartment to three men. They claimed they were police officers. Two of them were wearing masks; another was dressed in a black leather jacket. As they came in, they pulled out two guns, according to a police report. Poteyev himself and his son were tied up with tape. Was it the FSB having found out about his work for the Americans? Fortunately, the men were just robbers after money. And there was a lot in the apartment—they took more than three thousand dollars and thirty thousand rubles. That might in itself have been suspicious since it could raise questions about where it all came from (neighbors remember Poteyev frequently changing the foreign cars he owned and tipping the janitor five hundred rubles for helping him to his apartment when he had drunk too much). For now, though, Poteyev wasn't suspected of anything and could breathe.

NEARLY TWO YEARS later, Sergei Skripal's world came crashing down. When he had returned to Moscow from his posting in Spain, MI6 was faced with the same challenge as the Americans with Poteyev. Skripal was initially working in the GRU's personnel department and his family life appeared settled, with his wife, Ludmilla; a son, Sasha; and a daughter, Yulia. It was too risky to meet MI6 officers in Moscow. Instead his wife delivered a book with secret writing to British intelligence in Spain on two occasions. In 1999 Skripal quit the GRU, saying he was fed up with corruption. He went to work in the private sector for the last commander of Soviet forces in Afghanistan and continued to sporadically supply information to MI6, now traveling in person to Spain and Turkey for meetings. But the hunt for spies in Moscow was intensifying.

In December 2004, FSB officers pounced on him just outside his apartment. He knew better than to struggle against the men half his age. His shoulder was wrenched from its socket and a hood placed over his head before he was bundled into the back of a blacked-out van and driven off. TV cameras were there to film it all. He was taken to Lefortovo prison in Moscow—a site with a dark history and home to political prisoners and those accused of spying. His trial was held largely in secret. He was only shown to the cameras at the very end, looking confused but unbowed, wearing a track suit in the colors of the Russian flag as if to make one last claim of patriotism. Russian intelligence sources briefed that the sentence of thirteen years was lenient since he had cooperated—although whether that is true or how far is unclear.

Conditions in the labor camp were tough, but Skripal had been a championship boxer in the Soviet army (leaving him with a slightly squashed nose). This meant that even though he was now more than fifty years old, he could take the blows from guards and deter some of the other inmates from giving him the treatment they often dished out to political prisoners. Eating prison slop, he would dream it was ice cream. The reason for his arrest, it later emerged, was thought to be information provided to the Russians by a Spanish intelligence officer who was aware of his original recruitment in Madrid. Senior MI6 officers went out to Spain to take part in the investigation. There was a despondency in the organization. When an agent is caught, there is both the personal regret for an individual to whom they had promised safety and professional regret that they had been outplayed, making others less likely to follow his path. But this was not the only exposure of British intelligence operations in Moscow.

The embarrassing incident featured a rock—about the size of a football—that was heavy enough that it required two hands to lift. Picking up a rock in a snowy Moscow park and carrying it off is not exactly the most inconspicuous activity for a British diplomat. Nor is kicking the same rock. But in January 2006, Russian TV broadcast surveillance footage of four Britons amid sensational claims about a "spy rock." It was explained that an agent could walk past the rock and press a button on a handheld device kept in his pocket. That would transmit intelligence that would be stored in the rock until an MI6 officer later picked it up. It was the latest high-tech version of the classic dead drop, with the advantage that the agent did not even need to hide his material. "According to our experts, this device cost millions of pounds. It's a miracle of technology," an FSB spokesperson said. X-rays of the rock were shown on TV revealing its secret compartment along with footage of the alleged British spies (one of whom warily looks around as if suspecting a camera might be close and then slows down and glances at the rock as he passes by).

The revelations surrounding the rock were closely linked to a high-profile campaign. The TV documentary in which the footage was shown claimed that British spies were contacting and financially supporting Russian nongovernmental organizations (NGOs). Now a new law would crack down on them. An alliance of NGOs and foreign spies was portrayed as responsible for the so-called Color Revolutions—the "Rose Revolution" in Georgia in 2003 and "Orange Revolution" in Ukraine in 2004, in which pro-Russian leaders were removed from power. This, the Kremlin and the FSB thought, had been part of a plot. All the talk of promoting democracy—and the various groups involved in supporting its growth in Russia—was actually cover for subversion by Western intelligence. The conspiratorial worldview of those in power in Russia meant sometimes they connected dots and saw conspiracies and plots when there were just events. They were convinced that the guiding hand of Western "special services" (their term for intelligence services) was behind these events. The West was also edging NATO closer to what Russia defined as its "sphere of influence," with talk of Georgia and Ukraine one day joining. The sense of paranoia in Russia was growing. The need for intelligence, like that supplied by the illegals to uncover the intentions of its adversaries, was more important than ever. And so was the need to stop Western spies stealing Russian secrets and plotting within their country.

In 2007, the head of the FSB, Nikolai Patrushev, said that in the previous four years more than 270 active officers and 70 agents of foreign intelligence services had been uncovered. "One should specially single out Britain, whose special organs not only conduct intelligence in all areas but also seek to influence the development of the domestic political situation in our country," he explained when asked who was most active, citing the Skripal case. Foreign spies were doing more than just stealing secrets. "Giving themselves the credit for the disintegration of the USSR, they are now nurturing plans aimed at dismembering Russia," Patrushev told journalists. Russian spies were convinced the West was stirring up rebellion in the Caucasus. For Poteyev, watching the arrest of Skripal and the intensifying spy fever under Putin and Zhomov would have created a sense of rising pressure for him. How long could he manage without making a mistake?

Russia's spies were changing. In the 1990s, they would often talk nostalgically with their American and British counterparts about the Cold War. But by the mid-2000s, a new generation was taking over. They were colder and harder, more conscious of the humiliations of the past. One American recalls witnessing the tension at the end of a long vodka-drinking session after a liaison meeting. An older Russian officer reminisced wistfully about the good old days of the Cold War, when the two spy services went head-to-head. But a young FSB officer reacted angrily. The older officer's generation was the one that lost the Cold War, he said bitterly. His generation was determined to restore Russian pride and would take the fight to their enemy.

14

The Controller

On March 31, 2002, Richard Murphy was letting off steam to the man opposite him. The pair were at a restaurant in Sunnyside, Queens, in New York. The two men met there once a year. It could have been two old friends catching up, one listening to the other about the struggles of middle-aged life. Murphy was unhappy with the way his work was going, his career was not progressing as he had hoped, his wife was on his back, and he was short of cash. His dining partner, a few years older with a lean face, eventually ran out of patience. "Well, I'm so happy I'm not your handler," he said to Murphy. As they were finishing, the other man stood up and said there was forty in a bag. The two were not friends. Both were illegals. But they had different jobs.

The illegals were designed to be in deep cover with no contact with the Russian embassy. But they still needed human contact with Moscow. And Murphy's lunch partner was the key. He was a wily, veteran spy and one of the most important illegals—but also the only one who did not permanently reside inside the United States.

His name was Christopher Metsos. Except, of course, it was not. It was Douglas Cox. Or Sean O'Donaill. Or Diego Cadenilla Jose Antonio. Or it could be Patrick Woolcocks. The real

Patrick Woolcocks's brief brush with Russian espionage came as he emerged from a restaurant in Moscow in 2004. Woolcocks, a Briton from Hampshire, worked in sales for a company that supplied TV and video technology and he was in the city on a business trip. He and some colleagues had just enjoyed a meal together in a Scandinavian restaurant when he left in a taxi. But he only made it twenty yards before the car was stopped. Two men came up to him. They were dressed in leather overcoats—almost the stereotype of FSB officers. "Papers, papers," they demanded. He dutifully handed over his British passport. They took it away to their vehicle, parked nearby. The minutes ticked by. Eventually they returned and handed the passport back to him. He thought little more about the incident until years later, when MI5 would interview him and as he began to have trouble at border controls. His identity had been stolen.

In all, there were at least eleven identities. Those were just the ones the FBI knew about. Some sounded Irish, some sounded Latin American, some Russian. The man who used them was in his fifties, with light brown hair, and was balding. Sometimes there were glasses, sometimes there were not. Sometimes there was a mustache and sometimes there was not. He was around six foot, an expert in disguise and a black belt in martial arts. The only distinguishing marks—a scar on his chest and burn or pock marks on his arms—were going to be hard to spot from a distance as he traveled the world changing his name and his face. The best guess is that his real name was Pavel Kapustin. But he has become known as Christopher Metsos.

Who or what was he? Metsos was what is called a Special Reserve Officer of Directorate S, also known as a "Traveling Illegal," while the FBI called him an "Illegal of the Center." These are experienced career intelligence officers who have usually already done their time as deep-cover illegals in another country. After returning to Moscow they are assigned to work as part of a team run out of the Special Reserve Unit of Department 1 of Directorate S. They are older and more experienced than the younger couples and individuals. They can be used for short-term, often risky missions. If an American whom the Russians wanted to try to recruit was seen in Cairo, a Special Reserve officer might be sent out to pitch to him. The advantage was that if it failed and this illegal was reported to the American or Egyptian authorities, then they would have already disappeared. If a member of the *rezidentura* in the embassy had tried and failed, they would be identified and potentially expelled, and if a deep-cover illegal tried it and failed the whole investment in their cover would be wasted. These officers can also be used to attempt a false-flag recruitment—posing as someone from, say, Spain in order to recruit someone who might be willing to betray secrets to that country but not to Russia. It goes without saying that they needed not just multiple identities but the best documents so they could move quickly and securely across borders.

The central identity this man used was Canadian. Metsos used a real passport obtained in the name of a real child who had died at age five. He had at least ten dates of birth for the years he was supposedly born, ranging from 1954 to 1959, although it is doubtful he had a party for each birthday. The FBI believes he first came to the United States for operational work as early as 1993, according to documents, but they will not say how much they know of his early career. In 1994, the then thirty-six-year-old was studying at Norwich University in Northfield, Vermont, a military academy founded in 1819. The two thousand students are half military cadets and half civilians. Metsos said he had come from Bogotà, Colombia (although the address he gave did not exist and the phone number he used belonged to a car wash). After the first half of the term in his sports medicine major, he disappeared. One possibility is that he was there to try to recruit a student who was on his way to becoming an officer in the US military, another that he was building a long-term cover and something went wrong. Where he went next is a mystery, like so much about him.

Special Reserve officers had the task of supporting deep-cover illegals. Metsos was based in Moscow but traveled around the world to meet them, including those in the United States. The systems of covert communications were the means for regular messages and the passing of orders from Moscow and intelligence back from the United States. But illegals also needed money. And this was not so easy. Nothing could be more incriminating than a transfer from a Moscow bank account into that of one of the illegals. And even payments routed through various third countries would still be a red flag to any investigation. Cash was far harder to trace. But it required someone to deliver it. And that was Metsos.

On May 16, 2004, an FBI surveillance team was waiting and watching in Queens, New York. They had video cameras on multiple positions around the Forest Hills station on the Long Island Rail Road. The cameras were carefully hidden. Video surveillance in such a fixed location was easier and safer than trying to have people follow the suspects and was possible thanks to the FBI's source in Moscow.

The FBI was expecting a particularly significant meeting. Metsos and a second secretary (Poteyev's old job title) from the Russian Mission to the UN were both spotted at the station separately but carrying nearly identical orange bags. Metsos walked up the stairs. The Russian walked down the stairs. It was a narrow stairwell and they passed close by each other in the middle. As they did, each handed the other the orange bag they had been carrying. This was a "brush contact"—a quick handoff. Metsos kept walking up and the Russian down. In the briefest moment, an exchange had been done. The bag was thought to contain a quarter of a million dollars. The money had gone black—moving from official diplomatic channels into the world of espionage. And the cash was destined for the illegal network. In this case, two illegal families—one in New Jersey, the other in Seattle.

Just hours later, Metsos was seeing Richard Murphy at the Sunnyside restaurant for one of their regular meetings. The FBI had it wired for sound and pictures. Metsos handed over a package that, he said, contained Murphy's "cut." It is thought this was about half—around $125,000. Metsos explained that the rest of the money being handed over was separate from Murphy's cut. "You will meet this guy," Metsos explained to Murphy, "tell him Uncle Paul loves him. . . . He will know." There also seemed to be a bizarre recognition phrase. "He will know it is wonderful to be Santa Claus in May."

THE FBI HAD installed a GPS tracking device on Metsos's car. The next day it revealed him traveling eighty miles north to a rest stop near Wurtsboro, in Upstate New York. The FBI later found something in the area where the car had stopped. There was a partially buried upside-down brown beer bottle right next to a log and close to a telephone pole. The FBI team started to dig carefully. After they had cleared away five inches of dirt, they found a package wrapped in duct tape. A dead drop—a hiding place where something is left by one spy for another. The point of contact between two spies is the moment of greatest vulnerability and what a counterintelligence service is looking for. That is why a brush-past needs to be fast. A dead drop has the advantage of avoiding direct contact. The two spies will have been in the same place but separated by time.

Inside was the other half of the money. The FBI photographed it and left it in place. They set up a camera in a tree directly above to monitor this site. That way if anyone disturbed it, the FBI would have a good view. No one was sure how long they would have to wait to spot anything. The answer would be years. That was the way this investigation worked—long and slow.

Two weeks after the May 2004 Long Island Rail Road pickup, the FBI intercepted a phone call by Richard Murphy to Seattle, and to another pair of illegals. Mikhail Kutsik had arrived in 2001 under cover of being US citizen Michael Zottoli. The real Zottoli was born in Yonkers in 1970 but died a year and a half later. Natalia Pereverzeva arrived in 2003 as a Canadian named Patricia Mills. They were married on June 5, 2005, in Washington State. Both attended the University of Washington as business majors, graduating in 2006. One of his professors would occasionally share a shot of vodka with Zottoli and described him as one of the brightest students he taught. Washington State has always been high on the Russian target list, home to defense companies like Boeing, bases for nuclear submarines whose comings and goings can be watched, and communications infrastructure points as well as high-tech companies like Microsoft and Amazon. Zottoli and Mills's aim was likely to get into one of the tech businesses or build contacts with those working there.

On June 18, 2004, after the phone call with Murphy, Zottoli and Patricia Mills took a plane from Seattle to Newark airport and checked into a Manhattan hotel. The next day—at the agreed time of 3 p.m.—Zottoli arrived at one of the entrances to Central Park. An FBI team tailed him. Significant moments in the case may only come once or twice a year and two illegals meeting was one of them. They saw Mills sitting on a bench. She stayed there for an hour and a half looking toward the entrance where Zottoli was standing. She seemed to be carrying out countersurveillance for her partner, looking for anyone who might be tailing and watching Zottoli. She did not spot the FBI. But something went wrong. Murphy was nearby but Zottoli and Murphy do not appear to have spotted each other. Ironically, the FBI knew they were both nearby but the pair did not. The agreed location may not have been precise enough. After an hour and a half, Zottoli gave up and headed back to his hotel. That evening, Zottoli and Murphy spoke on the phone to try to work out what had gone wrong. "We might have, have different place in mind," Murphy said. "I was there at three." Zottoli replied, "I was there at three o'clock, too." The FBI listening in knew they were both telling the truth. They agreed to try again the next day.

At 4 p.m. the next day Zottoli arrived at the metal globe near the subway entrance at Columbus Circle. He was wearing chinos and a T-shirt. The FBI could also see Murphy, wearing a stripy T-shirt and shorts and with a backpack. Murphy was taking pictures—like a tourist. At one point he aimed his camera directly at an FBI camera taking pictures from above him. But it seems to have been just chance. The FBI caught the moment on camera that the two men met and shook hands. It was 4:01 p.m. They were smiling. The pair went into Central Park but after just three minutes they headed in different directions. Zottoli was now carrying a red bag with a museum logo that he did not have before. Zottoli had the money that Murphy had received from Metsos. This was the first time the FBI saw two US-based illegals meet. It had been a brief encounter—they were together for only five minutes, no time to share pleasantries.

Five days later, Zottoli and Mills headed back to Seattle and to their apartment. This set of meetings was hugely significant for the investigation. A trail had now emerged that went from a Russian official through Metsos to two illegal couples. But they would have to be patient to see what happened next. Two years later, on June 5, 2006, Zottoli and Mills flew to JFK Airport in New York. Three days after landing there they drove up to Wurtsboro. They were there to pick up the cash that Metsos had buried in Upstate New York in May 2004.

The FBI video cameras installed at the site caught Zottoli digging around the beer bottle and retrieving the package. The camera caught Mills nearby. The next day they were in Washington, DC. The FBI had the room wired. Zottoli was wearing a money belt as he left the hotel. When he was back the video caught him dividing money up among several wallets.

On April 17, 2005, Metsos and Murphy met again for one of their regular encounters at the Sunnyside restaurant. This time Metsos gave Murphy an ATM card and PIN so he could withdraw cash from an account funded by Moscow Center. The illegals as a whole had to be careful about the money they received from Moscow Center. Like any spy they had to avoid spending it flashily in a way that drew attention to them. But they also had to account for it carefully to their paymasters. Heathfield and Foley in Boston sent back itemized reports detailing how the money had been spent.

Christopher Metsos was much more than just a courier who handed over the money. When an intelligence officer from the CIA, MI6, or SVR recruits an agent—a citizen of another country who is going to betray secrets—they know that this person will require careful and subtle handling. What is their motivation? What are their fears? Understanding the difficulties of being an agent—and especially the isolation in not being able to tell anyone around you about it—is a key part of the job of an agent runner or handler. And even though they were not agents but intelligence officers themselves, the Russian illegals faced similar challenges in terms of isolation and strain. And they needed careful handling to keep them on track. That was the job of Christopher Metsos.

He was a mentor—there to support them, guide them, and watch over them on behalf of Moscow Center. He provided the human contact with their superiors that no coded email or radio message could offer. He was their lifeline back home. But he was also their controller. He would come out and deliver in person any message that the Center needed delivered. Sometimes those messages were supportive—keep going, even though it's tough, he could tell them. If they had complaints he could commiserate. And of course, he could say he knew what it was like to live such a life since he had been an illegal himself. But sometimes there were demands from Moscow Center for more intelligence and harder work from the illegals. Metsos could be the person to smooth that message—yes, Moscow can be demanding but you just need to learn to feed the machine, he could say. But at other times, he could play the bad cop and be firmer and more demanding—the Center expects more from you. He could give them a hug or give them a kick in the ass when they needed it, the FBI says.

"Traveling Illegals" like Metsos were the glue that kept the illegal operation running. Metsos met all of the family illegals based in the United States—but not always in America itself. He would meet some of them outside of the country. For instance Heathfield traveled globally for his work and both he and his wife could meet Metsos in Canada—this was safer since they all had Canadian identities. And it was not just the American illegals whom he met. As he crisscrossed the world on his false passports and stolen identities, he met deep-cover illegals dotted around the world. This made him an enormously important intelligence target for those covertly watching him. He carried secrets way beyond any of the illegals settled in the United States and could potentially unlock hidden aspects of the global program. But Metsos would not be able to travel to the United States for long. The illegals were about to be dealt a blow that would force them to shift gears and take more risks.

15

Murphy Steps Up

PAUL HAMPEL WAS about to board his flight out of Montreal at 6 p.m. on November 14, 2006, when things went wrong. Hampel was a well-traveled man with a business job that was so vague that no one who met him quite knew what it was. But it allowed him to visit Eastern Europe and the Balkans, and establish a corporate presence in Dublin. He even published a book of his photographs called *My Beautiful Balkans*. The Canadians who stopped him at the airport explained that he would not be catching his flight, as they had some questions they needed answering. It was when they searched his luggage that he knew he was in real trouble. He was carrying nearly eight thousand dollars in five different currencies, three phones, and a shortwave radio. He also had index cards with handwritten notes that included key dates in Canadian history, like that of the country's confederation, and also names of prime ministers and their parties since the 1940s. These were likely designed to help him if he was asked questions at border control on his return to test his "Canadianness." The truth was he had been operating for more than a decade as a deep-cover Special Reserve officer from Department 1 of Directorate S, with a particular focus on the Balkans. He was another Christopher Metsos. But while Metsos would travel into North America,

Hampel was based in Canada in order to travel to the Balkans to support deep-cover illegals settled there. Hampel, it would emerge, had received three successive passports since 1995 based on a fake 1971 Ontario birth certificate for someone supposedly born in 1965.

The arrest was the result of an investigation the FBI was involved in and perhaps even a tip from their source in Moscow. It spooked the Russians. The reason was that Christopher Metsos had been operating under Canadian documentation similar to Hampel's. Their fear in Moscow Center was that Metsos would now be identified if he tried to enter the United States. And if he went to meet the illegals, then they in turn would be blown. The SVR did not know that it was already too late.

The decision was taken in Moscow to stop using Metsos as an illegal traveler supporting the American-based illegals. He would never return to the United States after 2005. But the illegals still needed funding. That meant finding a new way of getting them cash. The new method would be high risk. Moscow Center would now deliver directly to one of the illegals rather than through the intermediary—known as a "cutout"—of Metsos. Murphy, the stay-at-home dad and unfulfilled spy, would be given the role of substitute courier now that Metsos was out of the picture. It was time for him to step up.

The FBI team believes that the illegals were given the new, riskier mission of carrying out the brush-pasts because Moscow thought that it was time for them to up their game. They were spies and they had been trained to take these kinds of risks and get them right. But one thing Moscow Center had not understood was the psychological impact created by the loss of Metsos. He had been more than just the moneyman. A veteran illegal himself, he knew when to reassure and when to cajole. Now the illegals had no one to talk to. The only contact with the Center was through the informal method of the covert emails. Imagine having a boss who never spoke to you face-to-face but only through emails. Some might prefer it, but it takes away the ability to have the informal conversations and read the body language that helps you know what your boss really thinks. Without the in-person catch-up with Metsos, the spy equivalent of a regular appraisal, some of the illegals, especially Richard Murphy, would struggle with the burden of their work. But first, Murphy had to show he had what it took to be the moneyman.

In April 2009, the FBI decrypted the first of a series of messages giving Murphy instructions to carry out a "flash meeting" with a "field station rep"—a Line N officer. Murphy would be taking over the role of picking up money and distributing it to other illegals. The exact sequences of events and the type of bag to be used were all detailed in the message. A Saturday had been chosen, Moscow Center said, because the commuter station was deserted and there were no surveillance cameras.

On June 6, 2009, the FBI had its own covert video surveillance all around the station. Murphy was spotted. He looked almost comic in a wide-brimmed, floppy hat and sunglasses and with a couple of different bags over his shoulder and in his hand. He was hanging out at the lower part of the station, occasionally checking his watch. A Russian government official (a third secretary at the UN mission) was seen on camera wearing a baseball cap and T-shirt, pacing on the platform. He was carrying a shopping bag as he descended the stairs. Murphy headed up the stairs. As they passed the midpoint, Murphy held out his backpack and the Russian placed a small satchel into Murphy's open backpack. Two FBI cameras caught the exchange, which lasted no more than a second. They did not speak to each other. But, perhaps because he was less experienced, it was still a messier transfer than the one Metsos had done at a train station a few years earlier. Murphy continued up the stairs to the train platform. In the bag was roughly $300,000 and a flash memory stick. Half the money was for him, half for others.

This meeting was crucial for the FBI investigation. Using Murphy as the contact man technically went against the strict rules of tradecraft, which were supposed to keep illegals distanced from each other and from Russia itself. The breakdown of the hard rules that had protected the illegals offered the chance to build a case. What was even better for the FBI was that it would soon have confirmation that the Russians had no idea they were being watched. Soon after, a message from Moscow Center to Murphy was intercepted. "Flash meeting: well done. A good job. . . . Our tech people in NY didn't notice anything suspicious." That message was met with joy and relief in the small FBI team. "We all high-fived ourselves because they didn't spot us," Derek Pieper of the FBI's New York team later recalled.

The Russians had their own technical team trying to monitor the FBI for unusual patterns of communications or activity around the meeting. But they had spotted nothing. In the world of counterintelligence, this was a real coup—the FBI listening in on the Russians talking about trying to watch the FBI to see if they had been watching two Russians.

After the train station meeting, the SVR communicated precise details for when and where Murphy should meet "Mike"—Michael Zottoli—to give him half the money and the flash drive.

On September 26, 2009, Murphy and Zottoli were due to meet on a street corner in Brooklyn. The FBI was waiting. But there would be one of the close calls that can define the success or failure of an operation. The FBI had multiple cameras watching the scene. One was in a car on the exact street corner where the Russians were due to meet. Zottoli, wearing a pair of jeans and with a laptop bag slung over his shoulder, was there first. The car was parked just a yard or two away from him. He was holding a magazine and glancing around, clearly looking for someone. At one point he looked straight at the surveillance car. Murphy, in a pair of white shorts and carrying a backpack, is then spotted and the camera pans as he crosses the street toward Zottoli. The pair stand a foot or so apart from each other, pretending they have just bumped into each other.

"Excuse me, did we meet in Bangkok in April last year?" Zottoli says. "I don't know about April, but I was in Thailand in May of that year," replies Murphy, exchanging their agreed code.

At that exact moment a man in a sweatshirt walked up to the FBI's surveillance car parked just feet away. The stranger pulls the door handle. He is trying to open it. This was a moment of complete fear. Were the Russians on to them? Was there a Russian countersurveillance team? The door handle rattled but did not open. The person walked off. Everyone breathed out. Was it a Russian? No, the FBI believes. It was, most likely, someone seeing if the car was unlocked and if he could steal it. "Welcome to New York," recalls FBI agent Derek Pieper with a smile. If the door had not been locked, then there would have been a potential grade-one disaster. It would have swung open, and if Murphy and Zottoli had seen a camera trained on them it would have been a calamity. "The case is over," explains Pieper. "The operation hinges on that door not opening. That's ten years right there."

Oblivious to the FBI camera two feet away, Zottoli and Murphy shook hands. There was something of a smile between them. And then something strange happened between the pair of illegals. This was supposed to be a brief meeting—more than a brush-past but only the chance to exchange a few recognition words and then hand over money, a few minutes at most. But it did not turn out that way. The FBI team watched in surprise as the pair of illegals headed off together on a walk. They went to nearby Fort Greene

Park in Brooklyn and sat down on a bench. An FBI team with a covert video camera followed them there. And the two began to talk. And they kept on talking and talking. For an hour and a half. This unexpected turn of events caused the FBI a problem. These were the days of videotape rather than digital cameras and they began to run out of tape. Derek Pieper had to literally run around to all the different FBI vehicles nearby to collect their blank tapes and deliver them to the team watching the bench. Later in the meeting, Murphy handed over something that he pulled from his rucksack and placed it into a smaller bag he had brought. The pair kept talking all the time and appeared to be smiling. Then they both stood up and walked away in separate directions.

The illegals were not supposed to meet each other unless they had to, in order to avoid compromise of one leading to another. But now the FBI had a flow of cash going from a Russian official to Murphy and then on to another illegal. They had also seen something against all the rules—a long conversation between two illegals.

The first in-person meeting between two illegals that the FBI observed had been Zottoli and Murphy in 2004. This lasted just five minutes. By 2009, they were sitting on a bench and talking for what seemed like hours. Why? Because they needed to talk. "It is that human nature thing—if I am annoyed at my boss, I turn to the guy next to me and say, Can you believe he did this? You have that ability to do that. But they had no one," explains Maria Ricci. "So human nature took over and what should have been a quick meeting ends up being a two-hour meeting because they couldn't help themselves and they didn't have Metsos to talk to. They didn't have somebody to sit down and say, These are my problems, please help me." On September 28, Murphy sent a message back to Moscow Center. "Meeting with M was successful. A passed to M the card and $150k." He did not mention the long chat.

In January 2010, Moscow Center began making plans for Murphy to travel back home to Russia the following month. One reason was to deal with some problems with communications. He was asked how long he planned to be back, bearing in mind that the amount of time required for "business" would be two to three days. Travel back to Russia was clearly sensitive and had to be carefully handled. The messages were precise about the dates and routes of travel through Europe. Brush-pasts had to be arranged in order to pass on documents along the route with a backup option in the event the first meeting was missed.

Murphy's cover story would be that he was either an interpreter or IT specialist on a business trip by invitation of a Russian businessman (he could choose which he preferred). A copy of an invitation letter would be enclosed in the envelope with the passport. The plan was for him first to go to Rome, Italy, for January 23 on a US passport. There he would carry out a flash meeting. The meeting place was code-named "Stan"—at the shop window of Libreria, near the San Giovanni subway station. The recognition signal was that Murphy was to be holding a copy of *Time* magazine so the title could be seen. If *Time* was in his left hand, then that would be a danger signal that something was wrong. The local Line N officer would say, "Excuse me, could we have met in Malta in 1999?" Murphy was to reply, "Yes, indeed, I was in La Valetta, but in 2000." Murphy was to be given an Irish passport in the name Eunan Doherty Gerard. This was the name of a real person from County Donegal. Gerard had been to Moscow on vacation in 2005 and had sent his passport to the Russian embassy in Dublin to obtain his tourist visa. That was most likely when the details were stolen.

Then it would be to Milan by train to catch a flight. He would be provided with a round-trip ticket the Center had booked using his Irish passport details. The e-ticket could be found in the drafts folder of an email address. The trip to Moscow was to update communications. Murphy was instructed to purchase a specific laptop made by ASUS before his departure. He was to buy it carefully—no ordering ahead, then paying cash and destroying the receipt. That computer then needed to be brought to Moscow.

On February 9, 2010, the FBI watched Murphy go into a Manhattan electronics store and come out with a laptop box. The FBI went into the store afterward and spoke to three employees. They identified Murphy. One remembered him saying his name was "David." They checked the sales database and found reference to a David Hillier having paid cash for the same ASUS laptop that Murphy had been told to buy. On February 21, Murphy took a Continental Airlines flight to Rome from Newark. Just before 9 a.m. on February 22 he made a purchase at the Rome airport as Richard Murphy. But a receipt found later at his house from lunchtime that day in Rome's train station was in the name of Doherty.

Murphy flew back from Rome on March 3—meaning he had perhaps a week at most in Moscow. What he did there was a mystery—but he may well have met with Poteyev. As he arrived back in Newark, Department of Homeland Security officials covertly searched his bags. The same make and model of laptop that he had bought in Manhattan was there. But it was not the same. There was a sticker on the bottom of the laptop. The one he had bought in Manhattan had a serial number ending 1432. The one coming back from Moscow ended 9719. The laptop had been switched. The SVR had been careful but not careful enough.

Now that he was back in the United States, the FBI watched Zottoli and Mills leave their Arlington apartment in their gray BMW sedan on March 6 at 12:25 p.m. By 6 p.m. they arrived at a hotel on the Upper East Side of Manhattan. The next morning, the FBI were in the Fort Greene neighborhood of Brooklyn, where Murphy and Zottoli had met in September 2009. At 11 a.m., agents saw Zottoli meet Murphy at a pay phone on the corner of Vanderbilt and DeKalb Avenues. Murphy had a backpack and Zottoli a duffel bag. Again it was supposed to be a quick meeting. But, like September, it would go on much longer. This time they went to a coffee shop rather than the park. No one had known they were going to that coffee shop (although it was always a possibility, given the weather was cold), but in one of the bizarre coincidences that sometimes characterize surveillance operations, inside sat FBI agent Maria Ricci. She had been sitting at a window seat for hours, ready to watch the two men meet directly across the street.

Now the two illegals sat down right next to her. She had been watching Murphy for years, knowing every detail of his life, while he had no idea who she was. It was almost surreal. But it was also an opportunity that could not be missed. Ricci texted another female on the team and organized for her to go to the women's bathroom with a surveillance camera. "I ended up doing my own brush-past in the bathroom to get a camera." She placed the small camera by her leg and tried to train it on the pair. This was not what she had expected to be doing. "God, I hope it's facing the right direction," she thought. The Russians again talked for much longer than strictly necessary. "They ended up talking for a couple of hours so I ended up sitting in that spot drinking coffee for four hours," says Ricci.

The FBI overheard the pair, who never spoke in Russian, discussing Mills's and Zottoli's problems with their computer equipment. "This should help," Murphy responded. He removed a plastic shopping bag with the laptop inside from his backpack and put it into Zottoli's duffel bag. It took a few moments and was far from the subtlest of handovers. "If this doesn't work we can meet again in six months," he went on. Then came a telling phrase. "They don't understand what we go through over here." The "they" was clearly Moscow Center. Ricci sat so close that when Zottoli leaned back in his chair after getting his backpack, his chair knocked hers and the camera shook. A second or two later, a figure came over and handed Ricci a coffee. Much to their later amusement, all three FBI agents had—by complete accident—worn exactly the same clothes that day: blue jeans and a black sweater. It was as if the message had gone out to tell them the FBI surveillance look for the day. Eventually, the long meeting was finished and the illegals departed.

Two days later, Murphy sent a message to Moscow Center reporting that the laptop and nine thousand dollars in cash had been successfully delivered. He also relayed the problems that Zottoli was having with the equipment, which involved some kind of "hanging" or "freezing." There was one more message to pass on. His wife, Patricia Mills, was worried about some fake travel documents she had been due to use. Zottoli said she could not leave the United States because the documentary requirements for entry at her destination had just changed that year. Zottoli and Mills needed advice on what to do about their cover problems.

The illegals were busy but they did not know that the clock was counting down on their life in America. They had just over three months left.

16

Anna Takes Manhattan

"RELENTLESS" IS HOW one of the acquaintances who encountered Anna Chapman in New York remembers her. He was a single, middle-aged corporate lawyer who enjoyed the good life. She was introduced to him at a fancy dinner in Soho. There were about twenty people there; more than half were women from various parts of the world. At first he thought she was just like the other young Russian women in New York who were all pretty and well dressed but very clear about what they were after. So he made sure she knew he was not interested in her type. But she would not give up. Somehow she got his number from a friend and began texting him, asking to meet. She kept texting. Again and again over the next few weeks. So he checked her out with a few people in the business community—well-placed people who had known her from her former life in London as well as now in New York, the type who worked in finance and traveled the Atlantic regularly, "hedge fund guys." They all vouched for her—clearly her time in London had been well spent; her connections were paying off. "She knew everyone," he says. Soon they were dating and she was around his place. It was not just the looks. "She had this confidence about her you don't see in many people. You could drop her anywhere and she would find her way," he says.

Later, when he found out the truth, he would wonder if she had had some kind of training in psychological manipulation. "She really understood people," he says, before pausing and adding, "men in particular."

From January 2010, Manhattan was home for Anna Chapman and she hit the town hard. She had been traveling back and forth from Russia for a while but now she was permanently based in New York. Her business card carried the slogan "Explore Your Possibilities." That was a pretty good description of what Anna Chapman was doing. She had worked the London party scene and now it was Manhattan's turn. She went to exclusive bars, wore the best clothes. There were rumors she was a millionaire. She was a superb networker—just like Donald Heathfield but in a very different world.

Her work was selling property in America to rich Russians online. A $1 million loan from a Russian government fund for start-ups may have helped her on her way. In April, she sat down for an interview for a New York Entrepreneur Week event. She had the start-up spiel down pat, explaining that the idea for the company had come about when she was trying to buy an apartment in London and could not find one place that brought together all the information online someone needed. And so she decided to start her own business "to help" people. She also claimed she had been an investment banker before giving that up to follow her dream in New York. "All dreams may come true if you act on it," she explained. "I was someone who just arrived in New York, I didn't know anyone," she said. "Now I know a lot of people who introduce me to someone else, and they introduce me to someone else." She was, staff recall, a great boss who was always a good gift giver and who managed to be both professional and enjoy life. A team member recalls that Chapman always wanted to keep the company servers in Russia.

Chapman was assiduous in looking for leads for people for her start-up but there was also the personal side. She spent a lot of time at high-end parties. "She was acting kind of scandalous," recalled one woman who had met Chapman at a club. "She was playing around, it was a joke, unbuttoning a guy's shirt. Not vulgar, but very flirtatious." She met hundreds and hundreds of people. Anna had one gift for a spy—she knew how to get men to talk. In six months in New York, she made serious inroads in terms of meeting influential people, mainly from the financial world rather than the political. The new breed of illegal was moving a lot faster than the old.

Bill Staniford met her soon after she started to visit New York. He was working for a property-tech company and some Russians he knew in the industry said they had a friend in town who was trying to start something up in a similar field. Could they meet? "She had no clue about real estate or real estate technology. None," Staniford recalls. But he was surprised at her fancy apartment at 20 Exchange Place near Wall Street, an art-deco masterpiece built in the early thirties. On the fifty-second floor, it had a killer view and left him wondering who was backing her (she would sometimes tell people she had Russian investors).

She was intelligent, engaging, and very gregarious, upbeat and full of energy, almost hyper, he found. She was certainly ambitious. "She wanted to maintain her lifestyle. That's what she really liked," Staniford says. Good restaurants, fine wine, and fancy clothes.

They did not go into business, but they did start a relationship. Staniford had an interesting past. A former marine, he was working as a military cryptologic linguist for the National Security Agency (NSA) in the early 1990s, holding top secret clearances. In Panama in 1992, he says he was approached by a Russian in a bar who seemed to know a lot of personal details about him.

Once Anna Chapman's real work was exposed, he would wonder if the Russians had targeted him because they suspected he was working for the CIA, since he was traveling frequently to Eastern Europe for his business (he was not doing anything secret). Even more bizarrely, his father had a longtime accountant who did his taxes. And who else worked for that small accountancy firm? Cynthia Murphy. If the Russians did want to know about him and his family, they would have plenty of personal details to work on. Even if he had not been targeted and it was all just coincidence, this was an example of how two different illegals might be able to offer two separate streams of information about someone—their finances and their personal life—and if they were of interest, then have Moscow Center use those to work out how they might approach them.

Chapman, Staniford says, would occasionally test him by saying negative things about the US government, criticizing its actions in Afghanistan. Chapman was also interested in meeting people—Staniford's cousin had been Speaker of the New York City Council. He also took her for a weekend to Las Vegas, where they walked around the Bellagio. "She was like a kid in a candy store." But he thinks Chapman realized he was neither CIA nor a particularly useful target for recruitment. The relationship continued in a casual fashion for the next few months, leaving Staniford one of those left in her wake after the arrest, like a whirlwind that had passed through. Another paramour of Anna's recalls a conversation where she opened up just a bit. She lamented that she might never have children or a normal family life. It was as if she knew there was a cost to the path she had chosen, however glamorous it seemed on the outside.

The FBI knew all about her love life because they had been watching her as soon as she had arrived in January 2010. The New York field office, including Derek Pieper and Maria Ricci, was on her tail. On January 20, an FBI team was at a coffee shop near the corner of Forty-Seventh Street and Eighth Avenue in Manhattan. They had covert video cameras for their surveillance mission. Seated near the window they could see the young redhead. She had a tote bag with her. Ten minutes after she arrived, a minivan passed by the window of the coffee shop.

Another FBI team was watching that vehicle. They knew it was being driven by a Russian official from the mission to the UN, Poteyev's former base. Chapman and the Russian never actually met or spoke but the FBI believed their proximity was no coincidence. This was the moment Anna Chapman was going to make contact with the SVR's Moscow Center, through a smart technological trick—a temporary private wireless network in which two laptop computers pair with each other.

One laptop is preconfigured to create its own private wireless local area network but only to communicate with another specific laptop based on its media access control (MAC) address. This is a unique address assigned by the manufacturer to a device that is publicly broadcast through a radio transmission when it is looking for a device connected to the internet. When the first laptop spotted the unique address of the second close by, it established a private network. Data could then flow between Chapman's laptop and that of the Russian outside. The advantage was that the data would never flow over the regular internet, where it could be swept up. There is obviously still a risk of that stream of data being intercepted by someone close by looking for it so it can also be scrambled using specialized encryption software.

It was a clever trick and should have been impossible to detect—unless of course you had a spy inside the SVR. Poteyev had told the CIA everything about Chapman and how she would be operating. And so the FBI team fired up a commercially available piece of software that searched for the presence of wireless networks.

They soon spotted one emerge between two computers. One was in the coffee shop. The other was in the minivan. It did not last long. It became clear the Chapman meetings were scheduled on a regular basis.

On March 17, a team of seven FBI special agents were watching Chapman. She was observed at 11:29 a.m. walking westbound on Warren Street between Church Street and West Broadway. She was wearing a yellow T-shirt and blue jeans and carrying a black bag. She went west down Greenwich Street and entered Barnes & Noble bookstore at 97 Warren Street at 11:35. She sat in the cafe and turned her laptop on. At the moment Chapman powered on her laptop, the FBI was able to detect a computer broadcasting a signal. Three minutes later, another computer joined it. One of the FBI team, Amit Kacchia-Patel, watched as Chapman plugged a cord into the laptop. It was not the power cord. The wire led into her bag. She began typing. Then she was seen pulling the cord out and putting it back in. Was something wrong? Meanwhile, the Russian was across the street carrying a briefcase. Chapman was in the store for half an hour. The Russian was outside for about twenty minutes. At 11:59 she left the table and three minutes later walked out of the store. At 12:20 surveillance was terminated.

Over the next six months, the FBI would see a clear pattern of activity. On ten Wednesdays they would observe Anna Chapman in the vicinity of the Russian government official. The FBI believes a variety of meeting points were used partly to test the communications system. How well would it work in a busy built-up city? Would the signal get through in Macy's, Saks Fifth Avenue, and Rockefeller Center, where there would be countless other invisible data streams flying through the air around it? But it was also about the SVR testing Anna Chapman herself. How good was the young woman at being a wannabe spy and using technology and tradecraft? What became clear was that Chapman was having problems with the signal. FBI agents would sometimes sit just feet away from her in a cafe and try to learn how she was using her spy computer. She would often switch the laptop on and off and plug the cord in and pull it out, apparently to try to get it working. The problems were not due to FBI interference. The FBI had wondered at one point about jamming or collecting the signals, but the legal technicalities of doing so were complex since you had to be sure not to affect any of the other myriad of signals going through the air in the same area. Watching these meetings was vital in collecting evidence to build a case against Chapman. Since she was not using a false identity like the family illegals, they needed to prove she was engaging in illicit behavior.

On April 17, disaster nearly struck. An FBI team was assigned to carry out surveillance on the Russian official who was going to communicate with Chapman. Normally the FBI did not need to actually follow him to the meeting. Because they were inside the communications of the illegals, they knew where the meeting was going to take place and so could simply stake out the location in advance and wait. One team would then watch Chapman and another the Russian. The FBI knew the Russians were using devices in their embassies and missions to try to monitor radio traffic. The content of any FBI radio communications would be encrypted, but if the Russians noticed an uptick in the amount of unreadable traffic when one of their diplomats left the embassy, then they might know something was up, and if they noticed that traffic move with the person, they would be highly suspicious that there was surveillance on the person. This all meant it was easier to just wait at the meeting point.

But surveillance resources were tight. Counterintelligence had to compete with counterterrorism. In some of the illegals' cases, surveillance teams stayed on a single target for years—the same team did the Murphy family for close to a decade because they did not need to be followed day-to-day or up close. But with the Chapman meetings, a team was required every two weeks. And because they were in close proximity to her in cafes, it had to be a new team every two weeks to avoid the risk of anyone being recognized. On April 17, a new team made a mistake. They followed the Russian diplomat as he embarked on his surveillance detection route when he left his office in midtown Manhattan. FBI agent Derek Pieper was sitting in position waiting for the meeting. When he realized what was going on he began to freak out. As he listened to the team move two hours early, he was screaming in his head, "STOP. STOP." But it was too late.

The Russian was due to take a two-hour dry-cleaning route before he met Chapman. But as he headed out, the official spotted the surveillance team (most likely by technical detection of the fact that signals were following him rather than visually seeing them). He immediately aborted his route and turned around to head back to the office. Meanwhile, the other FBI surveillance team saw that Chapman's laptop tried to communicate but failed to make contact. She wandered around for a while before leaving but returned to the site later (most likely at an agreed backup time).

This could have been a disaster. Had the operation been blown? The team held their breath. Only a trained intelligence officer would have spotted the surveillance and aborted. This was textbook spy behavior. It confirmed the Russian official was clearly no diplomat. But now a lot depended on how the Russian interpreted the surveillance. Since he worked out of the UN mission, the hope was that he might have thought he would be under occasional FBI surveillance whether they knew he was a spy or not, and this was simply part of that. The next few days were tense for the FBI team in New York as they waited to see what the fallout would be. Might there be a signal to Chapman that something was wrong?

To their immense relief, the next meeting took place on schedule. It would have been much more serious if Chapman had spotted surveillance, since she would not expect to have seen anyone, given she was under deeper cover. The advantage the FBI had with Chapman was that she was not a fully trained intelligence officer. The deep-cover family illegals like Heathfield and Foley had gone through rigorous training back in Moscow in surveillance detection routes, because if they had been spotted meeting with a Russian official it would have been a disaster. After all, they were supposed to be Canadian or American. For Chapman, living openly as a Russian, coming up with a cover story for meeting a Russian was easier and so the risk was lower. The nature of these new special agent illegals was that they did not receive the same level of training in tradecraft as the family variety. On the whole, the FBI team who tracked her every move was not impressed with Chapman's spy tradecraft skills. Where she was extraordinarily skilled was getting men to talk. That was what made her dangerous. Her potential, they realized, was in direct proportion to the willingness of men to talk to a pretty woman. And that, some of the female FBI agents reflected with a wry smile, was a lot. They watched in awe as she went to party after party—the kind of parties, Maria Ricci thought, she would not have the clothes to go to. They could see she was starting to get close to people who mattered. One FBI officer reckoned that another six months and she would have ended up being the illegal who might have gotten closest to real power and intelligence.

MIKHAIL SEMENKO WAS another of the new illegals. Whereas Chapman worked the New York business and party scene, Semenko targeted Washington. Boyish with short dark hair, he was book smart and spoke five languages but lacked the guile that made Chapman so dangerous. Born in 1982, he studied at Amur

State University, just a few miles from Russia's border with China. In 2003, he went to teach English in China for nearly two years. In 2005, he arrived at Seton Hall University in New Jersey. "He definitely didn't seem to be hiding anything," a former classmate later said. "He must have told me that he was Russian within two minutes of meeting him." He then worked in New York for a travel agency that organized high-end tours to Russia. Semenko followed the firm to Arlington, Virginia, outside of Washington, DC, in 2009. "He seemed a lot younger than 27," the man who rented him the apartment would recall—the deal sealed over some vodka shots. "He was a crummy spy and a complete slob, but such a nice kid," he would later say.

He began applying for jobs at Washington, DC, policy institutes. On his LinkedIn profile he described himself as having "in-depth knowledge of government policy research." His page on a Russian social networking site had a picture of him posing in front of the White House. A friend in 2008 had written jokingly (one can only assume) on his page, "Hi to our valiant spy deep behind the nasty Americans' lines. Remember the teachings of Mao: destroy the filthy imperialist economy from within!!" Semenko returned to Russia in late 2009 and April 2010, likely for debriefing from the SVR. Once he obtained permanent status, the FBI believed he would have moved deeper into Washington and could have done real damage by burrowing into policy-making circles rather than just spotting people who worked in them, like Donald Heathfield. Semenko had the advantage that he was not under a false name so he might have been able to apply for jobs with security clearances or get to know people holding one without fearing a background check. Like Anna Chapman, he was an avid networker, turning up to embassy and think tank events. Experts on China were impressed enough by him and his blog on the country's economy to put him into contact lists as a "Russia/

China" expert. "He was a smart kid," one FBI agent says. "His drawback was his personality—people didn't like him. He was a little arrogant." One person told the FBI that Semenko had the personality of Vladimir Putin—a compliment inside the SVR but not in America. It meant he got stuck at entry-level jobs in think tanks like the World Affairs Council. FBI agents say he found answering the phones below him—given his education and secret status as a spy.

At 11 a.m. on June 5, FBI agents were inside a restaurant on Wisconsin Avenue in Washington, DC, when Semenko entered. Ten minutes later, a car with diplomatic plates entered the parking lot. It was driven by the same Line N officer who had carried out the 2004 brush-past at the train station in Forest Hills and who had returned for a second posting. After twenty minutes sitting in the car, he drove off. A few minutes after Semenko departed the restaurant. A wireless network was detected of the same type Anna Chapman used.

Four days later, Semenko came up to James S. Robbins after an event and asked if there might be any openings at the American Foreign Policy Council (AFPC), where Robbins worked. The council was home to a number of noted Russia experts with close ties to government. Semenko mentioned his blog and handed over his business card. Even though Robbins had studied Soviet intelligence activities, he did not suspect Semenko. "He came across as friendly, bright and earnest, the very kind of young person one regularly encounters in these venues. In the film version, he could be played by Harry Potter star Daniel Radcliffe. But in that respect, he was like many American intelligence operatives I have met; they are trained to be likeable." Semenko emailed the president of the AFPC soon after. Semenko was taking a direct approach in the way Heathfield had never been allowed to.

The third of the "True Name Illegals" was Alexey Karetnikov.

In his early twenties, he arrived in the United States in October 2009 (most likely having spent some time in the States as an intern the previous year). What made him stand out was his job in America—he worked as a software tester at Microsoft headquarters in Redmond, Washington. He came across as somewhat naive about America—expressing surprise to a neighbor when he did not have to produce his passport to buy clothes at a local store. The relatively poor language skills and cultural understanding were a world away from the older family illegals who had spent years working out how to blend in. The FBI believe Karetnikov still had potential to cause long-term damage. Silicon Valley has long been a target for spies from around the world seeking America's most advanced technology. At the start of 2010, Google was the target of a highly sophisticated cyberattack by China that reached inside the company's systems looking for its source code. Karetnikov might have offered a different avenue for Russia to do something similar. Getting people inside one of the major tech companies would offer a huge advantage. The FBI informed Microsoft and the company conducted a detailed audit of the code and projects he had been involved in. They found nothing suspicious and put in place new controls to detect "insider threats." Twenty years down the line, though, Karetnikov could have maneuvered himself into an influential job. These new illegals were moving much faster than the old. They had been born out of necessity—the challenges of documentation—but also opportunity—the post–Cold War deepening of ties between Russia and the West, which allowed young people to travel over. Even though they were openly Russian, they were gravitating quickly into positions of influence in New York, Washington, and Silicon Valley. They were the new threat. And at the same time, the risk posed by the established deep-cover illegals was also growing.

17

Closing In

THE MURPHYS' PUBLIC life as Americans and their secret life as Russian spies were on the up but sometimes the two came into tension. In August 2008, they moved from Hoboken to Montclair, New Jersey. This was a step up in terms of neighborhood and property. Thirty-One Marquette Road was not a huge house, but its three bedrooms and garage made it more spacious than their old apartment. It looked the part—a colonial-style home with maroon shutters and a porch, and near a park, it signified the American dream. "They seemed to have taken a class in Suburbia 101," a local teenager later reflected. Neighbors thought the new family fit right in. "They couldn't have been spies," one neighbor said when the truth later emerged. After all, Cynthia liked to garden. "Look what she did with the hydrangeas."

A few people wondered how the strange accents sat with an Irish name. Cynthia told one person she was Belgian. Others thought the couple might be Scandinavian because their daughters—Katie and Lisa—were so blond. The girls were popular. They sometimes ran a lemonade and brownie stand to raise money for charity. They were often seen out on the street on their bikes or having water fights. At school, they were good at languages—something they may have inherited from their parents. Katie received a certificate for Spanish, while Lisa was starting to learn Mandarin. Katie would graduate at the top of her class in June 2010. "I was just struck at how accomplished she was," said a parent who attended the ceremony. "I remember they called her up to the stage and they said, 'Stay right here. You're getting more awards.' " Their certificates were pinned to the fridge door. The girls knew nothing of their parents' secret life. Their one trip to Russia had come when both girls were very young. It was a chance for grandparents to meet them, but the girls would not have to be told where they were going or why. Otherwise they were just enjoying an all-American childhood, unaware that they were being watched.

Richard Murphy liked to grill hamburgers while drinking cheap American beer in the backyard, often dressed in a baseball cap and jeans. Neighbors soon realized there was one thing slightly unusual about him—he was a stay-at-home dad. How modern, they thought. He would occasionally have a coffee with them, but some thought he looked a bit sour and unhappy. He was often seen looking after his backyard vegetable garden after Cynthia got the bus to Manhattan in the morning. Cynthia now seemed happier, no doubt because she was the one whose work—both cover and real—was proving more successful than his. This made her seem like a happy working mother to those around her. Not all the neighbors may have been what they seemed, though. At some point, the FBI moved their own team into a house on the road to watch the Murphys.

It was their new house that caused a problem with Moscow Center. When they were renting, the couple had argued that they should be able to own a property. The response was unsympathetic. Moscow Center replied that none other than the director of the SVR had personally decided that if they had to buy a house, it would be the Center that would own it and the illegals would

A rare photograph of Alexander Poteyev, one of the most important spies of the modern era, seen here as part of a special forces team in Afghanistan in the early 1980s *(standing, third from the left)*. He was the key source on Operation Ghost Stories. *Origin unknown*

Potoyev worked for the KGB and then its successor, the SVR, and was later recruited by US intelligence. *Origin unknown*

BELOW: The spy swap at the Vienna airport on July 9, 2010, with the Russian plane in the foreground and the American plane in the background. *Matthias Schrader/AP/Shutterstock*

The statue of Felix Dzerzhinsky being toppled outside KGB headquarters during the August 1991 coup in Moscow. *Roberto Koch/Contrasto Image Eyevine*

LEFT: Robert Hanssen, the FBI agent who spied for Soviet and then Russian intelligence services. *FBI*

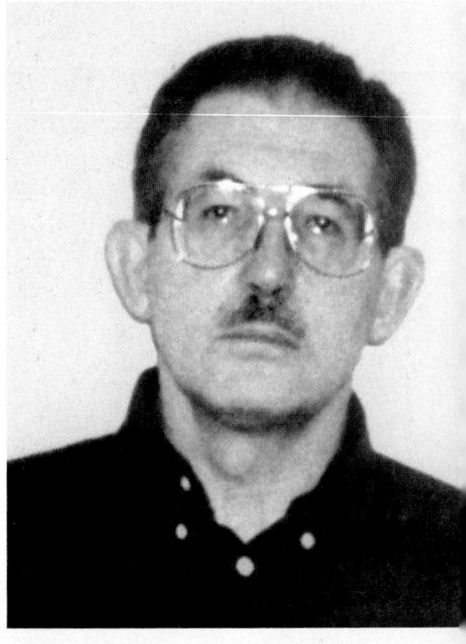

Aldrich Ames, the CIA officer who betrayed vital secrets from the agency's "Russia House" to his new friends in Moscow. *FBI*

President Obama convenes a meeting of senior national security staff in the White House Situation Room in October 2009. *Official White House Photo by Pete Souza*

Leon Panetta at CIA headquarters soon after becoming director in 2009. *CIA*

Presidents Obama and Medvedev enjoy a burger in Arlington, Virginia, on June 24, 2010, at the height of the "Russian reset." Three days later, the Russian illegals were arrested. *Official White House Photo by Pete Souza*

Milt Bearden, who ran the CIA's Soviet and Eastern Europe division at the end of the Cold War, pictured outside the Lubyanka KGB headquarters. *Courtesy Milt Bearden*

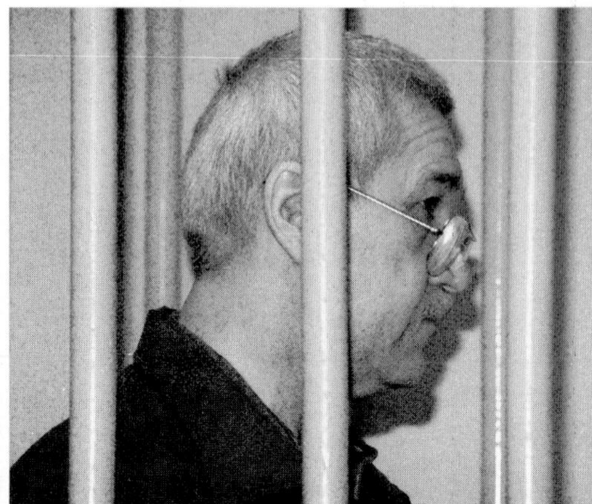

Alexander Zaporozhsky, the former KGB and SV officer who provided vital information to the CI pictured here at his trial in Russia, 2004. He wou be swapped in 2010 and travel back to the We *ITAR-TASS News Agency/Alamy Stock Photo*

Michael Sulick, a Russia specialist who rose to become head of the CIA's clandestine service at the time of the 2010 swap. *CIA*

A playful picture of Alexander "Sasha" Zhomov, the legendary KGB and FSB officer who led operations against the United States for decades. *Courtesy Milt Bearden*

Yasenevo, headquarters of the KGB's First Chief Directorate and later the SVR, and home to Directorate S, which ran the "deep-cover illegals." *SVR*

Yuri Drozdov, the former illegal who ran Directorate S in the closing decade of the Cold War. *Zuma/ Eyevine*

Vladimir Putin's employment ID while he was working with East Germany's Stasi spy service, dating from the late 1980s, when he was a young KGB officer. *BStU*

Ministerrat der Deutschen Demokratischen Republik
Ministerium für Staatssicherheit
Bezirksverwaltung
Dresden

B 217590 *

The illegal Donald Heathfield, aka Andrey Bezrukov, after his arrest in June 2010. *US Marshals Service*

An FBI surveillance picture of Ann Foley—real name Elena Vavilova—at her husband Donald Heathfield's graduation from Harvard in 2000. *FBI*

Donald Heathfield with his two sons, Timothy and Alex, in 1999. *Courtesy Alex Foley*

Ann Foley with her son Timothy at the Toronto Zoo. *Courtesy Alex Foley*

Brothers Alex and Timothy Foley in Bangkok in 2011, after their parents were revealed as Russian spies. *Courtesy Alex Foley*

The illegals Richard and Cynthia Murphy at home with their daughters. *Origin unknown*

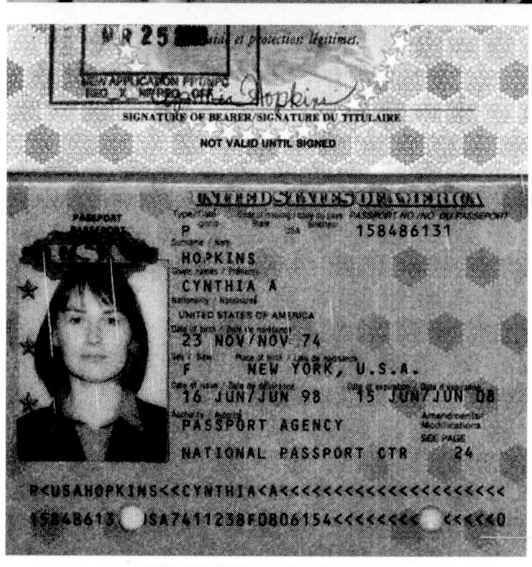

Cynthia Murphy's passport. *FBI*

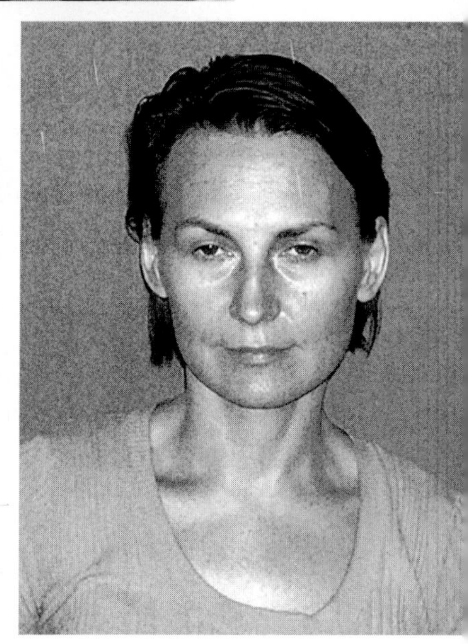

Cynthia Murphy after her arrest in 2010 *US Marshals Service*

Cynthia Murphy's Columbia University identity card. One of her tasks was spotting students who might later join the US intelligence community and become Russian targets. *FBI*

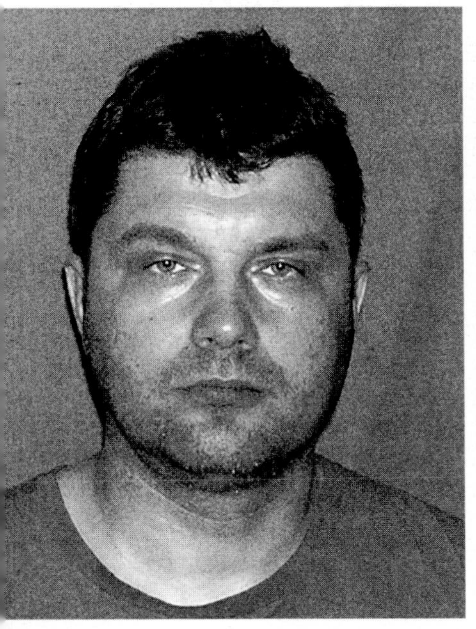

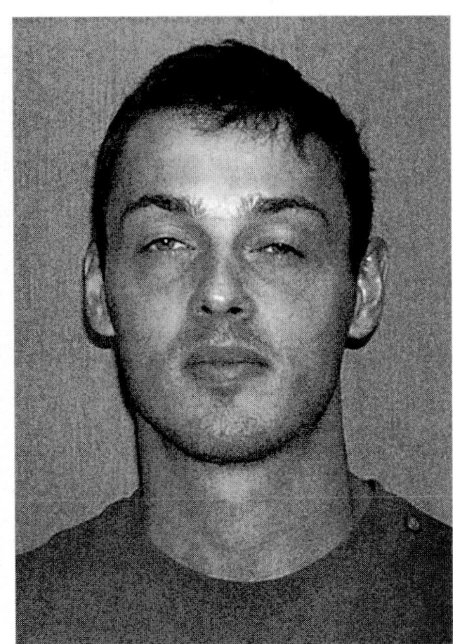

ichard Murphy *(left)* and Mikhail Semenko *(right)* after their arrests in June 2010.
S Marshals Service

Richard Murphy's notebook with the code to activate the secret steganography system used to communicate with Moscow Center.
FBI

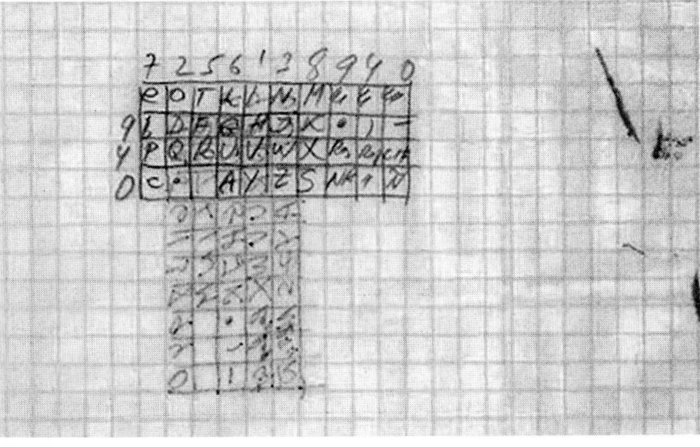

The checkerboard code Juan Lazaro used to send messages to the Russian Foreign Intelligence Service.
FBI

Christopher Metsos, the Russian spy who looked after the American-based illegals, seen here in July 2010 at the time of his arrest in Cyprus. Unfortunately, he would soon escape. *FBI*

Metsos used multiple identities during his time as a Special Reserve Officer of Directorate S. The date and locale of this photograph are not known. *FBI*

Another image of Metsos, this time in America, taken by FBI undercover surveillance in Brooklyn, New York. *FBI*

Christopher Metsos is caught by a covert camera in 2004 carrying out a brush pass, picking up money from a Russian official in a New York subway station. He would then give it to illegals working in the US. *FBI*

The marker left as a signal by Metsos in Wurtsboro, New York, where he buried money for the illegals. *FBI*

An FBI surveillance photograph of Richard Murphy meeting with Christopher Metsos in Queens, New York. *FBI*

The two illegals Michael Zottoli *(left)* and Richard Murphy *(right)* caught by FBI surveillance video, meeting on a Brooklyn, New York, street. *FBI*

Zottoli and Murphy talk in a café, not aware that FBI agent Maria Ricci is seated next to them. *FBI*

FBI surveillance video image of Murphy *(left)* and Zottoli *(right)* meeting June 20, 2004. *FBI*

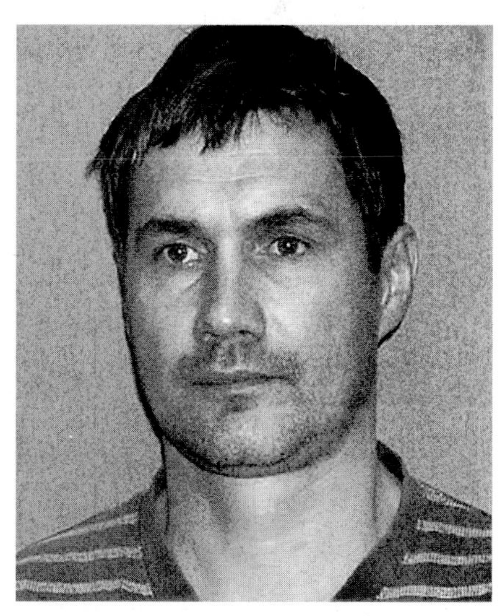

Zottoli after his arrest in 2010. He and his wife had been based on the West Coast but later moved east. *US Marshals Service*

A college identity card belonging to the veteran illegal Juan Lazaro. He was retired when he was arrested. *FBI*

Anna Chapman back in Russia after the 2010 swap. *WENN Rights Ltd/Alamy Stock Photo*

Anna Chapman in Manhattan before her arrest. *Polaris/Eyevine*

An FBI surveillance image of Anna Chapman on the street in New York. *FBI*

Former FSB officer Alexander Litvinenko and his wife, Marina. *Courtesy Marina Litvinenko*

Litvinenko in the hospital after he had been poisoned with radioactive polonium in a teapot in London. *Natasja Weitsz/Getty Images*

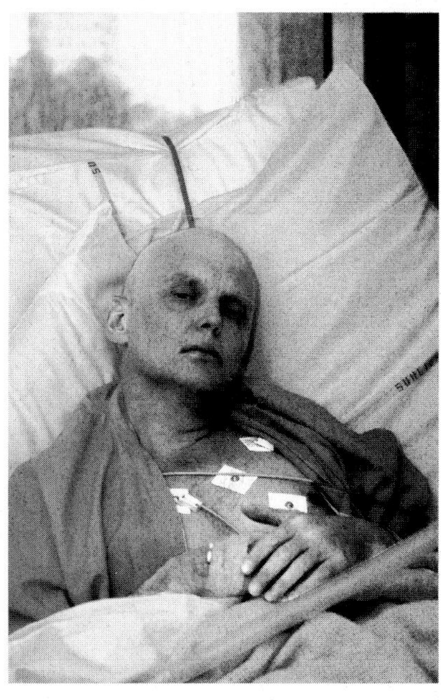

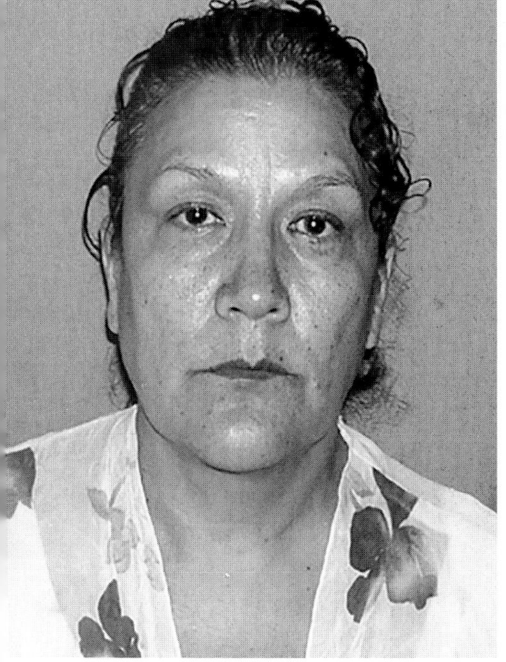

Vicky Pelaez after her 2010 arrest. She maintained she had not known that her husband, Juan Lazaro, was a Russian illegal. *US Marshals Service*

An image of Sergei Skripal as a young officer. After joining the Soviet military intelligence service, the GRU, he was recruited as an agent for Britain's MI6. *Courtesy friends of Skripal*

Skripal with his family in the mid-1980s. He is holding his daughter, Yulia, who would be poisoned along with him in Salisbury in the UK. *Courtesy friends of Skripal*

Maria Butina, a Russian alleged to have run back-channel communications to American political figures for Moscow. She has denied being a spy. *Polaris/Eyevine*

simply be permitted to live in it. The Murphys were annoyed and wrote a reply. "In order to preserve positive working relationship, we would not further contest your desire to own this house. We are under an impression that C [Center] views our ownership of the house as a deviation from the original purpose of the mission here. We'd like to assure you that we do remember what it is. From our perspective, purchase of the house was solely a natural progression of our prolonged stay here. It was a convenient way to solve the house issue, plus to 'do as the Romans do' in a society that values home ownership. . . . [W]e didn't forget that the house was bought under fictitious names." The Murphys were clearly trying to allay any fears that they had gone "native" in the materialistic American society. The purchase of a house had been part of their cover, they were saying. It was not a sign they had bought into the American dream.

The longer the illegals spent in America, the more embedded they were in American life. This made them at once more of a threat to the FBI, since their contacts were deepening, but also made the relationship with Moscow Center more distant. This was always the fear for the Center—that illegals would get too comfortable with Western habits and comforts and too isolated from their homeland. It was a fear with good foundation since, over the years, a number of illegals had found that the reality of life in the West did not match with the propaganda they had been fed. In 1980, one disillusioned KGB illegal in New York walked into the US embassy in Moscow while he was home and identified himself. But the KGB caught him and he suffered a false diagnosis of a mental illness and was given a ticket to a psychiatric hospital. In other cases, illegals returned from life abroad with heretical views about communism and their careers were cut short. The dissonance, heightened by not being able to talk to anyone about it, often caused something of an identity crisis for illegals over the decades. There might be tension in a marriage, affairs, or drinking problems.

The absence of the human connection that Metsos provided was another reason why arguments flared between the Murphys and the Center. The only way to let off steam was in the covert messaging system. This was helpful for the FBI team as they could see the anger build and more details of the Murphys' life being revealed than might otherwise have been the case.

The SVR sent a message during the argument over the house: "You were sent to USA for long-term service trip. Your education, bank accounts, car, house etc.—all these serve one goal: fulfil your main mission, i.e. to search and develop ties in policymaking circles in US and send intels [intelligence reports] to C[enter]."

The message was meant as a rebuke, reminding the Murphys why they were in America. But for the FBI it served as a neat, clear summary of what the mission of the illegals really was.

AFTER A DECADE or two of embedding themselves in American life, some of the illegals were beginning to get close to power. And politics was increasingly their target. A new administration arrived in January 2009 after the election of Barack Obama and the SVR was desperate to understand what it meant. A detailed "infotask" came to the Murphys in spring 2009. President Obama was about to make a high-profile visit to Russia. The SVR wanted details on the US position on plans for a new Strategic Arms Limitation Treaty, as well as on Afghanistan and Iran's nuclear program. Moscow said it "needs intels" on these topics and anything that related to key members of the Obama administration team who dealt with Russia, just below the cabinet level. Four specific individuals were named as people Moscow wanted to know more about.

Then comes one of the more telling instructions: "Try to outline their views and most important Obama's goals which he expects to achieve during summit in July and how does his team plan to do it (arguments, provisions, means of persuasion to 'lure' [Russia] into cooperation in US interests)." The SVR was starting from the position that the United States would be "luring" it into working with the United States and wanted to know how it would be done. The implication is that cooperation is some kind of trap for Moscow, which the SVR wants to be clever enough to spot and point out to the country's leaders. What they did not realize was that at this time, those members of the Obama national security team were genuinely hoping they could "reset" relations.

In another message, on October 18, Moscow asked the Murphys for more information on "current international affairs vital for R[ussia] highlighting US approach and providing us comments made by local experts (political, economic), scientist's community. Try to single out tidbits unknown publicly but revealed in private by sources close to State department, Government, major think tanks." On some occasions, the SVR indicated that intelligence was especially valuable. This was sometimes rather odd information. Cynthia Murphy sent back information about the global gold market, which the SVR seemed to think was particularly useful and was sent to two ministries in Moscow.

Moscow was aware of the limits placed on the Murphys by their fake documents. In May 2008, a decrypted message said that there were "three major ways [for Richard Murphy] to start [his] career for Service's purposes"—one of which was to get involved in "dem./rep campaing [*sic*] HQ in your area." A government job was hard but a political one was easier to get. In a 2009 message, it reminded Cynthia that "placing a job in Government (direct penetration into main object of interest) is not an option because of vulnerability of your vital records docs." Unable to carry out a "direct penetration," the next-best thing was to find people who in turn could be recruited as sources on behalf of the SVR and who could burrow in. This was why universities had long been the key place for Russians to talent-spot potential agents in the West.

Columbia University was a prime target. Cynthia began studying there in late 2008 for an Executive MBA and the SVR told her to collect information on her university associates. It is hard to recruit a serving Western intelligence officer unless they are particularly disaffected, but if you can recruit a student and then tell them to join an intelligence agency, you may have more chance. This is exactly what the SVR's forerunners did with Kim Philby and the Cambridge spies. So Cynthia was asked to "dig up" personal data of students. In particular, the Center was interested in students who had applied for a job at the CIA or who might do so in the future. And she found a remarkably easy way to do this.

The career fair at Columbia was full of stalls for prospective employers on the hunt for the most able students. There were the banks and the law firms offering the big money. But there were also tables for those who might be more public-service minded. One of those was for the CIA. People interested in the agency could sign up so they could be contacted. Cynthia Murphy simply watched her classmates who went up to the table and later noted down their names and sent them to Moscow. It was an incredibly simple thing to do and yet potentially enormously valuable for the SVR. It could be used to identify recruits who might end up going undercover and begin a file on them. They could either be spotted abroad or perhaps a personal weakness could be uncovered that might lead to recruiting a mole—a new Aldrich Ames—within the agency. The ideal for the SVR was a penetration agent—recruiting someone young and then directing them inside. "These aren't people who are valuable in any way, shape, or form today. But could they be ten years from now, fifteen years from now, twenty years from now? Absolutely," says Alan Kohler.

This was how the Russians worked—long term. And this was why, unchecked, the illegals posed such a risk.

The SVR directed Cynthia to "strengthen ties [with] classmates on daily basis incl. professors who can help in job search and who will have (or already have) access to secret info." She was told to report to the SVR detailed personal data, including "character traits," with "preliminary conclusions about their potential (vulnerability) to be recruited by Service." This is classic agent-spotting working. Anyone who looked good at Columbia would be run through the system back in Moscow to see if they were worth pursuing and having a file built up.

The Murphys did their best to deliver the goods by sending names from the university to the SVR, which then checked their database to see if a potential target was "clean." Sometimes the individuals came back with hits on the database. One contact of Cynthia's was reported to have been suspected by a then–Soviet bloc intelligence service of belonging to a foreign spy network. She was told to stay away and avoid deepening the contact for security reasons, since it could compromise her. In early 2010, she told Moscow Center she was interested in taking a job that involved lobbying with the US government and dealing with other foreign governments as well. But she was worried that the job might lead to an extended background check. Two days later, the Center replied that they had talked to the documentation department within S and "they don't see any hazards. . . . They . . . don't dig too deep during one's background check." They advised her to go for the lobbying job because "this position would expose her to perspective [*sic*] contacts and potential sources in US government." It was the kind of position that could draw her close to highly influential people.

There was head-scratching in some parts of the FBI, former officers say, at the way the illegals seemed interested in people who worked in the markets, academics, fund-raisers close to politicians, think tanks—and the like. The width and breadth of the cultivation was a surprise when set against the traditional view of Russian targeting, in which they focused on classified information. Some questioned whether the illegals were just "living off the government tit," as one former bureau official says, just enjoying life in America at the SVR's expense. But others pointed out that the Russians would not have continued funding the operation unless they felt they were getting some benefit. Only later would it be appreciated that their targeting signaled a shift in Russian behavior toward greater interest in politics and influence rather than just secrets.

On February 3, 2009, Cynthia reported a major success. She "had several work-related personal meetings with a prominent New York based financier and was assigned his account." The message said the financier was "prominent in politics," an "active fundraiser" for a political party, and a "personal friend" of a current cabinet official. This was Alan Patricof, a friend of Hillary Clinton, a fund-raiser for the Democratic National Committee (DNC) who also sat on the board of overseers for Columbia Business School. Hillary Clinton was one of the highest-priority targets for the SVR—she had just been appointed secretary of state. Anyone close to her could be of great benefit in picking up useful information.

Moscow responded that Patricof had been checked in the SVR's database and he was "clean"—with no signs of being a dangle. "Of course he is very interesting target," the SVR responded. "Try to build up little by little relations with him moving beyond just (work) framework. Maybe he can provide [Murphys] with remarks re: US foreign policy 'roumours' [*sic*] about White House internal kitchen, invite her to venues (major political party HQ in NYC for instance . . . etc.)." It ended with a clear sense of encouragement.

"In short, consider carefully all options in regard to" the financier. But Murphy was only ever able to work on Patricof's taxes rather than cultivate him in any deeper way. "She never once asked me about government, politics, or anything remotely close to that subject," he later said.

As well as Cynthia Murphy's attempts to cultivate people close to Hillary Clinton, there were other attempts by Moscow to try to gain influence around her. "In the end, some of this just comes down to what it always does in Washington: donations, lobbying, contracts and influence—even for Russia," Frank Figliuzzi, a former FBI assistant director for counterintelligence, later said. Such an influence campaign directed at a powerful figure with the potential to direct policy toward Russia should be no surprise. This is where the work of the illegals needs to be seen in a wider context. Their mission in getting close to officials and understanding the way in which power and influence flowed in Washington was part of a broader strategy. The illegals were just one cog in this bigger machine and the information they fed back could be used by other parts of the state. An illegal might identify a person who was well connected with a politician whom an oligarch could then approach with a business deal for instance. The Kremlin could coordinate between undercover illegals, state businesses, and oligarchs in order to secure influence. Cynthia Murphy's success in getting close to powerful people was starting to sound alarms. "We had seen enough of the reporting going back to Moscow Center to trouble us," Figliuzzi said soon after the case closed. "Several were getting close to high-ranking officials."

Michael Zottoli and Patricia Mills had also moved to the nation's capital. They had been struggling in Seattle. Despite his strong academic performance, Zottoli did not get the big job he wanted. He worked for a while as a car salesman and from July 2007, he was leaving at 7:30 a.m. every day to work as an accountant in Bellevue. Colleagues remember him as grumpy and unhappy. He was vague about his background and when people asked about his accent, he would respond with annoyance by saying, "Where do you think my accent is from?" He did not hide his politics. He occasionally went on tirades about President George W. Bush. Sometimes these would annoy coworkers so much that they would simply walk away. And it was not just the politics that was making him unhappy. Colleagues remember his phone would often ring. "My wife," he would say apologetically in his thick accent and then walk out of the office. Colleagues could still hear him argue with her on the phone. The feeling from what they could overhear was that he was henpecked or under the thumb.

In September 2008, Zottoli took six months off work, telling his employers he was going to see his wife's parents in South Africa (although she told people they were going to Europe). Zottoli was pushing Moscow Center to let him go to Washington, DC, where he felt he could make a fresh start and do more. Eventually, the SVR relented. In October 2009 the couple moved to Arlington, Virginia, just outside Washington, DC, with their two boys now in tow. The couple was not short of cash, with more than $100,000 in their joint checking account, but Zottoli had not produced much for all the investment placed in him by the SVR. Now he was told it was time to see what he could do in the nation's capital.

As they arrived in Washington, one of the things that worried the FBI was where they ended up living. There was always an element of fear when an illegal moved, in case they disappeared off the radar. But in this case the couple was in an apartment block just across the road from the Pentagon. And that meant it was packed full of people who worked in sensitive positions. In particular they lived on the same floor as people who had top security clearances. There were people in the building who worked in the Defense Intelligence Agency and even the FBI itself. That had the additional by-product of making life even harder for the bureau when it came to surveillance in the form of putting cameras and microphones in the apartment. These neighbors were the kind of people who might notice something odd going on next door. Zottoli and Mills were the least successful of the illegals, but they were still now in a position where they could start to make inroads. And you never knew which one of their neighbors they might have around for a beer and get friendly with.

DONALD HEATHFIELD WAS also making strides. There was talk of him moving down to Washington with his family. And the risks posed by his contacts were growing. One message indicates plans to build a spy ring at George Washington University: "Agree with your proposal to use 'Farmer' to start building network of students in DC," SVR Center said. Heathfield was playing the long game of penetration. Farmer—an unnamed individual but likely an academic—was perhaps going to recruit among students who could then apply for jobs inside the US government and even intelligence agencies (although Heathfield could have been exaggerating his cooperation). This, as with Murphy's work, was the real prize. In the past century people talked about the Cambridge spies in Britain; it could have been Columbia or GWU spies in this century if the illegals had succeeded. Ann Foley was also heading to Moscow for operational reasons. An SVR message gave her precise instructions: Paris to Vienna by train, where she would get hold of a fake British passport for a flight to Moscow the next day (she was told it was "very important" to sign her passport and practice the signature). She was told to "be aware" that she had just visited Russia and use a cover story about a business consultancy meeting on invitation of the Russian Chamber of Commerce. "In the passport you'll get a memo with recommendation. Pls, destroy the memo after reading. Be well."

• • •

THERE WAS ONE additional challenge for Heathfield and Foley—but one that might also offer an opportunity. And that was their children. They were older than the Murphys' girls and that posed a challenge that many illegal families faced. What did you tell them and when? Their parents had made a conscious decision that they wanted their children not to grow up simply as Canadians or Americans but with a more European outlook and a feel for the rest of the world. Was this just to reassure the parents and make their decision to mislead their kids easier—avoiding awkward patriotic dinner-table discussions of how wonderful America was and how bad Russia was—or was it perhaps because they feared one day their children might learn their secret? Or was it even because they themselves planned to tell them?

In 2008 Alex had become a US citizen as a result of his parents being naturalized. Timothy never took American nationality and always saw himself as Canadian. He was awarded first prize from the Canadian Consulate in Boston for artwork that promoted "Canadian values" and he returned a number of times, skiing with his family in Whistler and on road trips with friends (taking advantage of the fact the drinking age was lower over the border). The boys enjoyed video games and James Bond films (they had a full collection) even though Russians were not always presented very admirably. The brothers also occasionally played the Russian side in computer games set in World War II. "We were alarmed when in the end when their team won, the USSR national anthem was played," recalled their mother. The parents were no doubt disciplined enough not to stand to attention, but the music had a strange effect on them that they had to hide.

For any parent, deciding how to raise their children poses many challenges. Some parents will keep secrets from their children—perhaps a skeleton in the closet of their own past or that of their family. But what is it like for an illegal whose very identity—and therefore that of their children—is a lie that they have to hide? How long can you sustain the lie? What do you tell them and at what age? If you tell them too young, then will they go around the playground and boast, "My mommy or daddy is a spy"? What if you leave it too late? Your children are being brought up as Americans. What if they become Americans and when you finally tell them they decide they want to stay where they are rather than become a Russian?

These were discussions that the FBI could overhear many of the illegal parents struggling with. It was just one of the strange aspects of the investigation that the bugs would pick up these intimate parental discussions where the FBI knew the truth about the different parents but their own children did not. The parents worried that if they left it too late, their children would become Americanized. Would it be better to give up on the whole spying thing and go back home before it was too late? Or even try to stay in America and give up spying? The Murphys' children were too young to be told. But could they remain the Murphys and give up spying to become a normal family? Or do they give up spying and return to Russia and become the Guryevs again? "Each single one of them understood there would be a reckoning coming someday," says one FBI agent who watched the illegals. There was of course another option. Bring up your children with an eye to seeing if they might follow in the family business.

In Boston, Alex and Timothy's parents had wrestled for years with the question of what to do. One option was to retire to Russia when the children were young enough to adapt to life there. Another was to move to Europe as a kind of halfway house. They even wondered about returning to Russia but still under their cover as Heathfield and Foley rather than Bezrukov and Vavilov, which would allow them to go home but also to avoid having to confront the children with the fact they had lied to them. But in the end they had decided to keep working in America. Now the boys were growing up. That led to one of the most disputed elements of the case. Former FBI officials say the parents were definitely pondering whether or not children could be inducted into their secret life and become part of their operation.

This would not have been unprecedented. Rudolf Herrmann, an illegal code-named "DOUGLAS," was operating in America in the late 1960s and early 1970s as part of a husband-and-wife team (one of his missions was to penetrate the Hudson Institute think tank in Washington). He had been worried about his son growing up filled with anticommunist propaganda and had taken him on vacation to Europe to broaden his views. But he also saw the possibility of his son penetrating the US government or political life. During a trip to Latin America in 1972, he revealed his true identity to his teenage son, Peter, on a park bench. The son said he was ready to join the KGB and they went to Moscow, where Peter began training under the code name "Inheritor." Bizarrely, one of his classmates at high school remembers Peter actually showing him spy gadgets, including a secret camera and a coin that had a hidden space inside it. "This was a safe thing for him to do because whoever suspects a fifteen-year-old classmate to be a real spy?" the friend later recalled. Peter went on to study at Georgetown, where he was told to report on students whose parents had government jobs, especially if there was some vulnerability the KGB might be able to use, or provide details of fellow students who disliked American imperialism. He was also told to try to get a job at the Center for Strategic and International Studies, a think tank in Washington. But the father was confronted by the FBI and told to turn or else he and his family would be arrested. So there had been precedents in trying to turn whole families into operational illegals.

The risks from a new generation would have been significantly greater than that of their parents for the FBI. "It is that next generation which really would be damaging because they could pass a background check where [for] the traditional illegals, a deep background check would have uncovered some of the problems with the backstories," explains FBI agent Derek Pieper. "Their game is long and they will do it for a generation—and always the fear was when these people have kids and the kids grew up."

Timothy was inducted into the secret of his parents' espionage, according to US officials and Canadian documents. The Canadian Security and Intelligence Service informed Citizen and Immigration Canada (CIC) that Timothy was "sworn in" by the SVR prior to his parents' arrest. The implication is that this was a pledge of allegiance. Other sources suggest that this was more of an oath of secrecy never to reveal the truth of what his parents were doing.

Timothy attended George Washington University, full of people planning a career in government. Friends recall him being ambitious but with a good group of friends and who liked to spend time on the weekend at the Hawk 'n' Dove, a well-known bar off Capitol Hill. He studied international relations with a particular focus on Asia and he spent a semester in Beijing, learning Mandarin. In November 2009, he said he wanted to go into banking or "whatever makes me money." But in one account, the parents revealed to their son that they wanted him to follow in their footsteps. At the end of the discussion it is claimed he stood up and saluted Mother Russia and agreed to go to Moscow for training.

Timothy has always vehemently denied he was inducted into his parents' secret. "Why would a kid who grew up his whole life believing himself to be Canadian decide to risk life in prison for a country he had never been to nor had any ties to? Furthermore, why would my parents take a similar risk in telling their teenage son their identities?" He said the claim that he saluted Mother Russia is "just as ridiculous as it sounds."

Numerous officials have suggested that Timothy knew something. "It's logical to presume, and we suspect he knew something toward the end, before their arrests," Richard DesLauriers, who ran the Boston field office, said soon after the case became public. However, he did not suggest the same of his brother, Alex. "I'd say we have no reason to believe the younger son was witting of his parents' involvement."

AS THE SUMMER of 2010 approached, Donald Heathfield was planning to go to Moscow with his sons. "My brother and I had already been to most areas of the world. Russia stood out as a gap in our global coverage and so as a result of a year or two of pressure from us, my parents finally agreed to book a trip and apply for a visa," Alex said.

It was only in hindsight that some events from that June began to look a little odd to Heathfield and Foley. Her car had been taken away to a garage for a day. They had been trying to sell it and someone had supposedly wanted to have a look before buying it but then changed their mind. Later they would think that it was an excuse for bugging and tracking equipment to have been installed. And then there was a woman who introduced herself to Heathfield under some strange pretext in his work circles. She seemed uncomfortable, almost nervous, talking to him. And finally, someone in the neighborhood had popped around to their house—supposedly to look at how it was being refurbished. All of these events, the couple would later believe, were signs that the net had been closing. But of course if they had reported it back to Moscow, then there was someone there who could always reassure them that there was nothing to worry about, even though he knew the truth, and make sure no further action was taken. The net that had been around them for a decade was about to close.

• • •

EARLY IN 2010, FBI director Robert Mueller went to Langley to brief CIA director Leon Panetta and his senior team about the illegals. Even though the agency's Russia House had been involved in running Poteyev, the CIA had only limited visibility of the details of the FBI investigation that resulted from his work. Panetta, an experienced Washington hand, had been in office for a year but had known nothing of the Ghost Stories investigation until he was briefed on it, because it was so highly classified. The FBI team played videos and provided full details of the tradecraft. "It was a very important reminder of what Russia was trying to do to us," says Panetta, "also a reminder for us that we could not let our guard down." The illegals were becoming more deeply established. The decision was to keep watching them closely to see what they would do next. "We felt like we were really ahead of the game," says Panetta.

The first formal brief to White House national security staff—although not yet President Obama—came in February. The FBI explained the broad contours of the illegals program and some of the individuals operating in the United States. The case had been simmering for years. But it was about to come to the boil. And the decision on how to end it risked causing an explosion within the administration. But after a decade of patient surveillance, why was it now time to act?

The explanations as to why the investigation moved to its final phase have been confused, often deliberately to obscure the truth. False leads abounded. There were claims that the surveillance operation was somehow compromised. That was not true. There were also reports that the roundup was because the CIA wanted a swap to get their agents out and needed something to exchange for them. That got things the wrong way around. There was talk that perhaps Richard Murphy was being recalled because of poor performance or the illegals' term was coming to an end. But that was not the case. He might be frustrated and arguing with his boss, but his wife was doing pretty well. The official story at the time was that in spring the FBI learned that one or more of the agents were preparing to leave the country and might not return, and so if the opportunity passed, they would lose the chance to capture them. There was some truth in this. But it was not the whole story.

Another explanation was that the illegals were changing behavior and becoming more dangerous. This was a factor. There had been changes from 2009. That was partly the natural progression of an illegal's career—many had spent years building their cover and working their way into influential circles and this was now beginning to pay off in terms of the contacts they were making. As the summer of 2010 began, orders were coming from Moscow for the illegals to become more aggressive. They also seemed to be converging on the nation's capital. Donald Heathfield was making inroads. Cynthia Murphy was looking at lobbying jobs. Anna Chapman was blazing her trail through New York. Zottoli and Mills had been moved to the Washington area and along with Semenko were in physical proximity to people with top security clearances. The SVR was pressing for more results. They needed to prove their worth. There had always been some questions from the Center for the illegals to earn their keep and prove they were not just enjoying life in America. This meant that some of the constraints that had limited their activity—like how close they got to government officials or people with clearances—were now being relaxed. The risks were growing and some of the contacts were becoming more alarming. "We were becoming very concerned they were getting close enough to a sitting US cabinet member that we thought we could no longer allow this to continue," FBI officer Frank Figliuzzi later said. Cynthia Murphy's contact was one step away from a cabinet member—and not just any cabinet member but the secretary of state, Hillary Clinton. FBI officials were aware that if it later emerged that they had known a Russian spy was getting close to her and they had simply watched, there could have been severe criticism. "It would be withering," says one former official. The fears were growing. "Had they been allowed to continue, it's hard to say where their efforts would have ended," FBI director Robert Mueller said a few years later.

"We were getting intelligence and making sure they were not," is how one FBI official puts it. The bureau had got what they needed in terms of understanding the way the SVR worked. But it was getting harder to prevent bad things happening as the illegals embedded themselves deeper into American society and made more and more contacts. This was all true, but it was still not quite the full story.

Officials had to be careful at the time not to reveal that there was another factor driving decisions. Events were also driven by the man who had started it all. Ghost Stories had begun because of Alexander Poteyev. And he would bring it to a close.

18

Decision Time

IN MOSCOW, ALEXANDER Poteyev was growing nervous. A decade of clandestine meetings and the passing of secrets shredded your nerves. He had become worried the spy-catchers were on his tail. The relentless spy fever led by Putin and instituted by men like Alexander Zhomov at the FSB had been growing around him, the news filled with talk of arrests. And Poteyev had his own reasons to worry. He had been an enormously productive asset for the West. He had known the details of all illegals across the Americas and even though Directorate S was heavily compartmented, he had also picked up leads about other illegals operating globally (including the role of people like Metsos supporting them). He had shared these with his American handlers. As time had passed, Poteyev's intelligence proved so important that it was used widely and not just in the United States by the FBI. Some was shared with allies in a carefully protected way. This was because he had insight into illegals operating in their countries who posed serious national security risks. And this meant that in some cases action had to be taken.

The full tally of which cases were linked to Poteyev remains hazy, officials unwilling to confirm his role and cautious of revealing too much. But there was one case—one of the most important involving illegals in the 2000s—whose discovery set off alarm bells in Moscow and whose link to Poteyev has not previously been disclosed.

Herman Simm had just bought some cake at a shopping center and was walking back to his car when he was surrounded on September 19, 2008. The cake was for his stepmother and he had driven with his wife to a shop just outside the Estonian capital, Tallinn, to get it. Simm, just over sixty years old with a tough, rugged face that suggested he had seen much, was an important man. At one point, he had been head of the Baltic state's police force but more recently had been a senior security officer assigned by Estonia to NATO. That meant he was privy to some of the alliance's most important secrets. He was also a long-term SVR agent who had been inflicting massive damage on NATO. He had handed over thousands of pages of secrets and had seen almost all the traffic between the European Union (EU) and NATO, including details on secret codes used by the alliance. NATO officials said the damage was comparable to what Aldrich Ames inflicted on the CIA.

Simm had been tracked down through his handler—an SVR officer called Sergei Yakovlev who worked under illegal cover in Spain posing as a Portuguese businessman called Antonio de Jesus Amurett Graf. He was running a network in the Baltics. Simm had been placing memory sticks in trash cans to be picked up by the Russians as well as meeting Yakovlev in different countries. A call was intercepted between the pair in September, three days before Simm's arrest. At the time it was reported the investigation had begun in May 2008 and that the United States had been involved. The capture of a prized agent like Simm was a catastrophe for Moscow and the Russians began to investigate. One possibility, they would have known, was a compromise in Directorate S. When an operation goes bad, it immediately starts ringing bells. After all, it took a spy to catch a spy. Was there a mole?

The CIA had been careful to disguise leads from its agent when it shared them—the classic trick is to mask such material as signals intelligence, an intercepted communication. Some of the roll-ups over the years may have come from other sources or for other reasons but they would still have served to increase the focus on possible treachery inside Directorate S. Poteyev would have known that. The hunt for spies that had gathered pace in Putin's Russia was now focusing within the SVR's most secret directorate. After Simm, it would intensify. There would be talk in Russia later that their investigators were somehow zeroing in on Poteyev and that he was about to be polygraphed, which made him run, but that sounds more like an attempt to cover up failures in not spotting him. The truth was the pressure was starting to tell in the mind of the agent. If Poteyev was overly worried or paranoid, who can blame him? He felt the fear that every agent and their handlers know can come. He was done. After a decade, it was time to get out.

If Poteyev wanted out, then that would change everything. The success of Ghost Stories had been built on the fact the FBI and CIA were inside the SVR's operation. This had allowed them to know the identities, read the communications, and understand the plans of the illegals. It also provided the FBI with the confidence that there was no risk to sensitive classified material. But when their source was gone, they would no longer have that insight or control over the illegals. It was time to bring things to an end.

The FBI had begun to think about how Ghost Stories might finish as early as 2006. What happens if the investigation was compromised? What if Moscow either pushed the illegals to do something more dramatic or pulled them back? The FBI wanted the investigation to end on their terms. They did not want to lose the suspects by having them flee or just confront them and then allow them to walk away. After discussion, it had been agreed that the preference was to try to arrest the illegals. So the FBI had been building a detailed evidential case to get ready for this day. Word had gone out to the field offices to start collecting material with that in mind. Now they pushed this plan into action. But not everyone would be sure this was the right path. And some of the objections would come from the top.

Craig Fair was the operational lead in headquarters for the arrests. It was a delicate, unprecedented operation that was heading right up to the White House Situation Room. "It was one of the longer years in my career," he later reflected drily. Bill Evanina had been brought in to headquarters in 2009 to help. He had worked in New Jersey on the Murphy case but also had experience of the process of arresting and charging from time working on organized crime (after he helped advise the writers of the TV series *The Sopranos,* a character was named after him). Arrest plans were complex because of the number of individuals across different states and because field offices were not always aware of the interconnections. In spring of 2010 word went out to the field office that the case might be about to move to its final phase. Prosecutions were rare in counterintelligence since investigations normally relied on sources and methods that needed to be protected. This was going to be different. Evidence had been collected for years with the family illegals. There was the opportunity to send a message and put the Russians in the dock.

The initial time frame was that the end would come in the second half of 2010. But in May, the bureau found it did not have as long as it thought. It seems as if Poteyev wanted to get out fast. He thought his time was up and he had made that clear to his handlers. Top national security officials would tell cabinet officials that the source now needed to get out of Russia "immediately."

The whole operation was going to come to a conclusion not in months but in thirty days.

The exact timing was fiendishly complicated. Getting Poteyev out was a CIA operation and there was an exfiltration plan that had to be activated. But there were also time pressures because of the movements of the illegals themselves. The important constraint was that some of them were about to leave the country—including Donald Heathfield and one of his sons, who were due to go to Moscow on June 27, with the other son following soon after. Anna Chapman was also due to head back for a visit.

"There were a number of reasons there—not the least of which is several of the individuals were on their way out of the country and we would have lost our opportunity to detain them," Robert Mueller later said. This created a hard stop. If the arrest and escape date slipped beyond June 27 then the FBI risked having some of their targets out of reach. Not only was there a narrow window in which all the illegals were still in America but there was another prize that was suddenly, tantalizingly in reach. Christopher Metsos—who had not been seen in the United States since 2006—arrived in Cyprus on a Canadian passport on June 17. He checked into a two-star, forty-euros-a-night hotel just behind the waterfront in Larnaca. He behaved just like any other middle-aged tourist and looked the part as well, with shorts, an untucked shirt, glasses, and a mustache. An attractive woman in her early thirties with short brown hair would wait in the lobby and accompany him to the beach or dinner. Metsos's movements had been tracked back in Washington. Getting hold of a controller would be a major catch because his knowledge extended well beyond just the American illegals.

If this window closed, it might not open again. So logic said to move ahead of Heathfield's departure. Poteyev's departure had to be timed carefully to take place before any arrests as the Russians would instantly be aware of the possibility of a compromise in their own ranks. "It would clearly raise an awareness on the part of the Russians that there must be a potential source that made us aware," Panetta recalls. "In the intelligence business, when an operation goes bad it immediately starts ringing bells."

But there was a problem. And it was a big one. At the end of June, Russia's president was coming to town. And the White House was in the middle of trying to organize a high-profile "reset" of its relationship with Moscow. After a decade of patient intelligence work, the politics of diplomacy were about to risk everything.

OBAMA HAD ARRIVED in the White House just a few months after a new president took over in Russia. Putin was limited to two terms and Dimitri Medvedev had been handpicked to replace him. Younger, not just in appearance but in mind-set, than Putin, he certainly seemed to strike a different tone. Could there be a fresh start? In March 2009, Hillary Clinton sought to express this idea by bringing a prop when she met her Russian counterpart in Geneva. It was a large button. Unfortunately, thanks to a mistake it said "overload" rather than "reset" in Russian. And now there would be a fear that the arrest and prosecution of nearly a dozen Russian spies could overload the relationship at a critical moment.

There were different views about the "reset" across government. Inside the White House, the Obama team wanted big wins in foreign policy. Improved relations with Russia was one of those—a new START nuclear treaty was concluded in April 2010 and the two presidents drank champagne to celebrate. But it was not just about relations with Russia for their own sake but also the promise that this could unlock other issues. The debates over arresting the illegals were coming at a particularly delicate moment. In June, the Obama team felt they had scored a major diplomatic triumph with the UN Security Council approving new sanctions on Iran over its nuclear program. That had only been possible with the support of Moscow, which had not deployed its veto. Another aspiration—unsaid in public—was that they could boost Medvedev and make it more likely he could solidify a transition to a more moderate Russia. In Moscow, Putin was watching this new relationship carefully. He and other hard-liners were worried that their new President Medvedev was getting sucked in too close to the Americans.

Over at the State Department, Russia hands were skeptical Putin had really relinquished power but thought there was no harm in testing how much had really changed. Meanwhile, inside the CIA, old Moscow hands were much more critical, referring to the "reset bullshit" and those who believed in it as "naive morons." After all, they pointed out, in the summer of 2008 Russia, under its new president, had engaged in a small war with Georgia.

IN THE OVAL Office on June 11, 2010, President Obama and his top White House aides were handed a major headache by the FBI. The president was briefed by officials from the FBI, CIA, and Justice Department for the first time about Ghost Stories. He was given an overview of the illegals program, what the individuals in the United States had done over the last decade, and why events were moving apace. The possible criminal charges were outlined.

The FBI and CIA advocated tough action. These spies were called illegals for a reason, the FBI argued. The bureau had spent years on this case and did not want it to end in silence. Some in the FBI and CIA wanted to not just prosecute and imprison the Russians but go even further by also expelling dozens of Russian diplomats and other officials who were spying under diplomatic cover, to make the point. White House officials were not happy with the stance of their law enforcement and intelligence community. Medvedev was coming to the White House in a matter of weeks. The visit was being meticulously planned in order to give the impression of two leaders who were taking their countries forward. The last thing they wanted was a Cold War throwback getting in the way. Medvedev was going to travel across the United States before going to a G20 summit in neighboring Canada. The visit was due to last until June 27—the exact end of the arrest window. Arresting a bunch of Russian spies was going to be seen as a slap in the face. The timing could not be worse. "The primary concern was at that point whether this could perhaps undermine this effort to be able to work with the Russians because it would be so embarrassing," recalls then–CIA director Panetta.

The first White House meeting made it clear that there were major differences of opinion. There was a tangle of issues that were dauntingly intertwined. There was an anxious source in Moscow who had to be extracted from the country. There were ten illegals in America who needed to be arrested as soon as possible after his departure and some of them had plans to travel abroad. And finally, there were politics and diplomacy.

Mike McFaul was the Russia lead on the National Security Council. He knew Russia well and could understand the risk the illegals posed. "As I looked round the White House, I saw lots of young special assistants with the highest levels of security clearances. . . . Imagine if one of those illegals landed a job as an executive assistant to the national security advisor?" But the case, he felt, was also a reminder that there were those opposed to improving relations. And they were on both sides, he reflected. In Moscow there were the hard-liners who were watching Medvedev and hoping he would fail. But also he saw little sign of real belief in a reset for those in American intelligence and counterintelligence. "American Cold War thinking and habits were not going to change overnight," he thought.

White House officials—including the president—did not want to risk derailing relations with Russia. They did not think it was worth interrupting cooperation with Russia on other issues because of the illegals. The deliberation went on for days and was "heated." The resistance from political appointees to the CIA and FBI plan was intense.

The most aggressive option—imprisonment and a mass expulsion of Russian diplomats—was blocked by White House officials. Some at the FBI and CIA were "unhappy with what they considered a soft response," McFaul later conceded. "They wanted the court drama . . . they wanted the convictions. But Obama did not." There was—even more worryingly for the intelligence community—the real possibility that the whole arrest plan could be pulled.

On the afternoon of June 18, senior national security officials gathered in the Situation Room in the basement of the West Wing, taking their normal places around the long wooden conference table, with the president at the top. The politicians and diplomats had walked in angry and there was palpable tension in the air. The Medvedev visit was now just days away. There was still a clear division about what to do with the illegals. The political players from the White House and the diplomats were worried about the impact on the reset. Was it really worth taking the illegals down given the possible diplomatic fallout? Wasn't it going to look too provocative? Tom Donilon, one of the more skeptical, asked if there was a risk the operation would undermine Medvedev.

On the other side were the intelligence and law enforcement players from the CIA and FBI who wanted to get Poteyev out and bust the illegals. Robert Mueller briefed everyone on plans for the arrests and Panetta spoke about getting the source out. Obama was unhappy. "The president seemed as angry at Mueller for wanting to arrest the illegals and at Panetta for wanting to exfiltrate the source from Moscow as he was at the Russians," Robert Gates later said. "Just as we're getting on track with the Russians, this?" he recalls Obama saying. "This is a throwback to the Cold War. This is right out of John le Carré. We put START, Iran, the whole relationship with Russia at risk for this kind of thing?" The references were telling. In the minds of a new generation of leaders, all this spy stuff seemed a throwback and not the kind of thing that should get in the way of diplomacy. The view was that a bunch of dinosaurs on both sides who could not give up on old-school spying were risking an important breakthrough.

The case was made that the whole arrest plan should be shelved. Vice President Biden, sitting as usual to the president's right, was adamant that US national security interests would be best served by doing nothing at all, according to Defense Secretary Gates, recalling that the vice president argued that "our national security interest balance tips heavily to not creating a flap" that would "blow up the relationship with the Russians." Panetta, Gates says, disagreed with Biden and said that the real risk was the president being seen not to take Russian spying seriously. Looking weak was actually the danger. Hillary Clinton took a robust position in pressing for arrests. Others were willing to entertain the idea of arrests but only when the summit was done. But the problem was the source and the travel of the illegals. Obama himself was worried. His preference was to wait until after the summit to make the arrests, if they were made.

One option raised by White House staff was whether Poteyev's exfiltration could be held off until September. Risking a source was unacceptable for the intelligence officials. Panetta's priority was getting Poteyev out safely and he argued the arrests had to go forward. He had briefed defense secretary Robert Gates in advance. He knew that as a former CIA director, Gates would understand and support him. The two men made the case that they did not need to worry so much about the impact of the arrests. "This was part of a long story," Panetta recalls saying. "This is the way they do business, this is the way we do business." Each spied on the other. Sometimes you found their spies, sometimes they found yours. Diplomacy would continue because you can walk and chew gum at the same time. The reset would survive. But there was still resistance from the White House to the arrests.

What about letting the illegals go quietly back to Russia? That was a very real option that some wanted to pursue. But the counter to that was to argue there was a political risk. Panetta and Mueller pointed out that a small group of senior members of Congress had been briefed. "They were aware of what was going on with the spies and it would really be a serious political problem if Congress found out that we knew about these spies and did not take action to go after them and arrest them," says Panetta. "I think that argument carried a lot of weight." Panetta pointed out that if the story leaked, Obama would look weak and the Republicans might use it as a reason to not ratify the new START treaty. Obama also gave what some describe as a cynical but realistic answer, which was that if they let the illegals go, people in the CIA and FBI would be so angry they would leak details. "The Republicans would beat me up," Obama said. "Isn't there a more elegant solution?"

Gates weighed in on the side of the CIA and FBI in keeping the exfiltration plan on track, but he wondered if Obama could give details of the illegals to Medvedev and tell him to recall them within forty-eight hours or else face a much noisier expulsion. It might have even given him some leverage against Putin and the hard-liners who ran the program, he suggested. Some of those involved in the arrests believed they could use the arrests to embarrass Putin but the White House priority was on not embarrassing Medvedev.

Initially, Obama backed this option of talking to Medvedev

quietly and effectively letting the illegals go but discussions continued as the meeting broke up. Donilon was still unhappy with the plans and talked to Panetta. The top officials discussed things further and decided that confronting Medvedev would put the Russian president on the spot too much. The exfiltration followed by arrest was now the preferred plan. But they agreed they would try to do it in a way that did not humiliate Medvedev or rub the Russian's nose in the dirt. And that meant no prosecutions. "That would be long, exhaustive, and obviously even more embarrassing for the Russians," says Panetta. That left another option—a swap. There was still concern that this could take up to a year to negotiate, leaving a source of ongoing friction. But the aim was to "minimize the fallout," Panetta later recalled. "Russia would be spared the embarrassing spectacle of multiple criminal trials for their sleeper agents." There would be no prosecution. And—crucially—the arrests would have to wait until the Russian president had left North American airspace on June 27, to minimize embarrassment. "That was the decision and obviously it raised concerns about being able to retrieve—to protect—our source," says Panetta.

A consensus had finally been reached. The president himself was, according to a person involved, very unhappy but eventually came around to understand that it had to happen. Gates believed that the first instincts of the president and vice president had been "to sweep the whole thing under the rug" but he thought they had yielded to a wiser path. Some in the intelligence community still believed it was not enough. But a decision had been painfully made. Now it was time to execute it, and the timing was going to be tight.

19

Escape

ON JUNE 24, 2010, the patrons of an Arlington, Virginia, burger joint became eyewitnesses to the lengths the Obama administration was willing to go to in order to improve relations with Moscow. The president of the United States and the president of the Russian Federation both strolled in, jackets off, and began shaking hands with diners before ordering some food (which Obama insisted on paying for). True, they were not alone. There was a motorcade outside and Secret Service agents and a few journalists inside with them, but it was still not the place you would normally expect to see two world leaders who had just finished talking global affairs in the Oval Office. The pair sat at a small, cramped table almost knee to knee, and with their interpreters on each side. Strangely, Medvedev seemed to squirt a huge amount of mustard onto his bun but then eat the burger without it. The burger visit—with its studied informality—had been carefully choreographed by the White House.

The Russian leader's entire visit to the United States had been designed to present a new dynamic image of Medvedev. "Hello everyone, I'm at Twitter and sending my first message," the Russian leader had written while visiting the company's Silicon Valley headquarters a few days earlier in an open-necked shirt. Steve

Jobs had given him an as-yet-unreleased iPhone 4, while then governor Arnold Schwarzenegger reminisced that he had enjoyed a great time in Moscow back in 1987 to film *Red Heat*. "I'll be back," Medvedev joked with Arnie as he shook his hand, using the actor's catchphrase from the movie *The Terminator*. This, officials on both sides agreed, was what the relationship between Russia and the United States should be like—looking to the future and mutual cooperation, not stuck in the past with stories about spies and the Cold War. The two presidents were effusive at their closing press conference about the "new partnership" between their countries. This partnership, President Obama explained, was not just about government-to-government contacts but people-to-people ties. It was meant sincerely but what Obama knew and Medvedev did not was that the FBI was about to bust wide open one set of Russian "people-to-people" contacts that involved deep-cover spies.

As the two leaders munched on their burgers, Russian illegal Mikhail Semenko was living undercover about five hundred yards away, not realizing he was enjoying his final hours of American freedom. Meanwhile in Moscow, Alexander Poteyev was enjoying his last supper in his home country and making his final preparations to flee. On Friday the twenty-fifth he walked out of Yasenevo. One account says he rushed from a meeting, another that he booked some leave and threw a party for some of his SVR colleagues with good whisky, telling them he would see them soon. Another person says Poteyev actually left personal notes for some of his colleagues. He knew that either he would escape or get caught. Either way his time was up.

This was a moment of high tension. Once the plan had been set in motion, there was no way back. The arrests were scheduled in the United States on Sunday night, which meant that on Monday morning it would feel like a bomb had gone off inside Directorate S and Yasenevo. And the first question in the aftermath would be who had planted it.

THE ADVANTAGE OF Poteyev fleeing on Friday was the weekend ahead of him, leaving him two full days to get out. He headed to Tverskaya Zastava Square, home of the Moscow-Belorussky station, one of the major train terminals in the east of Russia's capital. There he boarded a train to Minsk, the capital of Belarus. His travel would not in itself be suspicious since he was originally from the former Soviet state. He told his wife, who accompanied him to the station, that he was going on a business trip (some say his marriage was not in the best state). As the train pulled out of Moscow on Friday evening, Poteyev would have known he could be seeing the city he had called home for the last time. That was unless he was caught in the coming hours. In that case, the next time he saw the skyline it would likely be glimpsed through the window of a prison van. The 434-mile journey would take the best part of nine hours. The wheels were now in motion and the escape had begun.

On the Saturday, as he woke in Minsk, Poteyev's wife received a text message. She would later show it to the SVR to prove she had no idea what her husband had planned.

It read: "Try to take this calmly: I am leaving not for a short time but forever. I did not want this but I had to. I am starting a new life. I shall try to help the children. Please do not turn them against me." The cat was now potentially out of the bag. Would the SVR realize something was amiss? The problem was that if they did, the FBI could not necessarily do much about it if the illegals were given an escape signal.

The Saturday morning began the craziest two days of the whole ten-year operation, the FBI team recalls. Poteyev had fled Moscow and would be missed on Monday morning. The Russian president was only flying out of North America on Sunday evening and the White House had insisted that no arrests be made before he was gone. There was the narrowest of windows to coordinate ten arrests across the country. Anna Chapman's was the one that came closest to going wrong.

THE FBI HAD a plan to entrap Chapman based around her "Wednesday" meetings using a "false flag" operation—having an intelligence officer pose as a spy from another country. In this case, an undercover American FBI officer was going to pose as a member of the SVR based at the Russian consulate, named Roman.

The use of the undercover was going to be important since Chapman could not be charged with fraud in the same way as other illegals because she had always used her true name. The plan was to charge her with failing to register as an agent of a foreign government. What was needed was evidence of her acting as one by following instructions given to her by someone posing as a foreign official.

At 11 a.m. on Saturday, June 26, Roman called Anna Chapman. He spoke Russian and said he needed to meet her urgently that day to hand something over. He used a code name that identified him as working on behalf of Directorate S and which would trigger a meeting with Chapman. A different name would have been an emergency signal to run. Knowing these coded phrases (known as "paroles") was vital to the operation. Illegals were told to obey orders from anyone who used the right one and only a few people knew the names, among them Poteyev. An hour and a half later, Chapman called back using the number he had given her. She thought she was calling the consulate, but she was not. The two spoke Russian again. It would be difficult to meet that day. She was in Connecticut. Could they meet tomorrow instead?

Roman explained it was urgent but said they could meet the next morning if they had to. The delay was bad news. The FBI could see the clock ticking down. But luckily Chapman clearly thought better of brushing off someone from Moscow Center and changed her plans. She called back half an hour later to say she would indeed come back to New York for the meeting. They agreed to meet at 4 p.m. at a coffee shop in downtown Manhattan.

The meeting started late. The undercover was wearing a blue shirt and chinos and carrying a plastic bag. He had a hidden recording device on him and there was video surveillance inside the cafe. Chapman was wearing jeans, a white T-shirt, and sunglasses.

He identified himself as the person who had spoken with Chapman on the phone earlier. "Do you want something?" he asked, meaning a drink. "I guess we should," replied Chapman. There was a brief discussion about who was paying and then they went to sit down. The undercover suggested she take a seat. It was in good view of the video surveillance team.

The pair spoke Russian initially, until the undercover suggested it might be better to talk English so they would draw less attention to themselves.

"How are you doing?" the undercover asked.

"Everything is cool," Chapman replied. "Apart from my connection"—apparently a reference to the technical difficulties she was having in the laptop-to-laptop communications.

"I just need to get some more information about you before I can talk," she said. She had not met this Russian before and she was suspicious.

"Sure," he said. "I work in the same department as you, but I work here in consulate. . . . There is a situation that I need your help with tomorrow, which is why it's not like regular email contact or website contact and this could not wait until your Wednesdays."

He thought it best to show attention to the laptop problems. Chapman was going back to Moscow in two weeks. He could take the problematic laptop back to the consulate to look at or she could take it back to Moscow. Chapman replied it would be more convenient to give it to the consulate. She rummaged around in her bag for a few moments. The undercover had a sip of his drink and she placed the silver Toshiba laptop on the table. Was it the encryption or the sending of the message that was the problem, the undercover asked? She could not receive messages, she explained.

It was time to lay the trap.

"I have a task for you to do tomorrow," the undercover said.

Chapman did not sound keen. "A short task?" she asked.

"Tomorrow at eleven o'clock in the morning. Can you do it?" he said.

Chapman audibly sighed. Being a secret agent could clearly be trying—especially for a busy young woman in Manhattan.

The man went on. "I will explain what it is but I need you to do it tomorrow morning."

"Okay, how long will it take because I need to like explain . . ." It sounded like she might have a date.

"Half an hour maybe." After that, he promised she could go back to her regular Wednesday schedule.

"This is not like the Wednesdays with the notebooks, this is different. It is, it is the next step. You are ready for the next step. Okay?" He wanted her to understand this was her chance to prove herself.

"Okay," said Chapman.

"There is a person here who is just like you, okay. But unlike you, this person is not here under her real name. . . . I have the documents for you to give to her tomorrow morning."

Was Chapman convinced? All she said was "Okay."

Chapman was curious. "Is she in New York?"

The undercover did not answer but tried to keep the pressure on.

"Are you ready for this step?" he said. Careful thinking had gone into how to handle Chapman. The strategy was to play on what they believed was her desire to be moved deeper into espionage work. It seemed to be succeeding.

"Shit, of course," she replied.

She needed to be at a bench at 11 a.m. the next day near the World Financial Center. Chapman would hand over a passport with a fake name to the woman. Now, the undercover handed over the document. He showed her a picture. "She will come to you; give her the passport and you are done."

There would be a recognition signal for the meeting. Chapman was given a magazine and told she had to hold it in her hand in a specific way. The woman would say: "Excuse me, but haven't we met in California last summer?" Chapman had to respond, "No, I think it was the Hamptons." There was a city map outside the cafe where she sat now. She had to place a postal stamp upside down on the side of the map while looking at it. He would then come and check it afterward. The stamp's presence would mean everything had gone okay.

The undercover asked Chapman to go over all the instructions. He added one thing he had forgotten, which was that she had to remind the recipient to sign the passport. Chapman complained she was tired and would have to explain all her travel that day. She was worried by her explanations to her boyfriend as to why on short notice she had had to travel three hours and would have to go back the next day. But she said she would carry out reconnaissance of the meeting place that evening.

She was nervous. "You're positive no one is watching?" she asked the undercover.

"You know how long it took me to get here? Three hours," he explained. That was a reference to how long the undercover was claiming he had been "dry-cleaning" in order to be sure he was not followed. The meeting was coming to an end. "Your colleagues back in Moscow, they know you are doing a good job and they will tell you this when they see you. So keep it up." Again, it was the attempt to play on Chapman's ego and desire to impress.

She was finding it tough and there was a moment toward the end of the conversation where she reached out. "Is it difficult for you?" she asked.

"This is my job, you know? It's busy, it's long days because I have to do work at the consulate and other work," he explained.

"So you work in the consulate?" she asked.

For four years, he explained. She was showing signs of suspicion. "So who instructed you to do this?" Chapman asked.

"I don't have any answers. I just have instructions," he said.

"Okay, it's just really scary," she said, suddenly sounding vulnerable. But she also seemed to sense something was not quite right.

The undercover tried to encourage her. "Good luck," he said. "I may see you in Moscow." That was unlikely. It was now around 5 p.m.

ON THE SAME day that Anna Chapman had been lured into a trap in New York, Mikhail Semenko was walking into his own in Washington, DC. Another FBI undercover posing as a Russian called Semenko on his cell phone. "Could we have met in Beijing in 2004?" the undercover asked, using an agreed name and recognition code. "Yes, we might have, but I believe it was in Harbin," Semenko replied. The two arranged to meet later near the intersection of Tenth Street and H Street NW in Washington, DC, about a five-minute walk from the FBI's headquarters. Semenko was two minutes early. At exactly 7:30 p.m. the FBI agent walked up to him. He was wearing a wire. The same recognition phrase was exchanged and they walked to a nearby park.

The FBI man explained he wanted to talk about an attempted electronic communication on June 5. The undercover said it had not worked. "I got mine," Semenko responded, explaining that everything looked fine. This was a problem for the FBI—the cover story was not being bought. Who had trained him to use the equipment? "The Center guys," Semenko explained. That was what the FBI wanted to hear. Was Semenko keeping the equipment safe? Semenko got annoyed and explained he would erase the hard drive if something went wrong. What about the prearranged meeting place when he received a particular signal? Semenko explained that his only meeting place was the Russian consulate in New York. That was evidence tying him to the Russian state.

There was a problem, though. Semenko seemed to be suspicious. He was curious about the fact that the Washington street corner they had just met at had been proposed as a potential meeting site but had never been approved. The FBI did not know that the SVR had actually rejected the location—perhaps because it was so close to the FBI. There had been a meeting to decide that recently. Only later the Russians would realize that one person had been absent from the meeting where that had been decided—Alexander Poteyev. The undercover handed Semenko a folded newspaper. This was almost as clichéd as you get in terms of spying: two strangers meeting and talking on a park bench and one handing over a newspaper with something inside. Inside was an envelope containing $5,000. The undercover explained that it had to be delivered to a park in Arlington the next day between 11 and 11:30 a.m. Semenko asked for a precise description of the location and was given a map. There was a spot underneath a bridge where the money was to be left. He was told to memorize the information on the map and then hand it back so it could be destroyed. But as he left, the FBI was wondering: had Semenko smelled a rat?

Meanwhile, things were going wrong with Anna Chapman. An FBI team watched her as she left the cafe. She headed to Brooklyn. At 6 p.m., they saw her enter a CVS pharmacy store. Next she went to a Verizon store nearby. This was odd. The FBI team began to worry. After leaving the Verizon store she went to a Rite Aid pharmacy and then went back to the Verizon store. This was now looking like seriously unusual behavior. She should have either carried out reconnaissance of the next day's meeting site or headed home. It also looked like she was doing a surveillance detection route. She had some training and was doing her best but she was no professional spy. But why the Verizon store? Was she buying a burner phone to make an emergency call? "I was kind of proud of her a little bit because this was her first well-executed operational act," says Maria Ricci, who was watching. "She went to a strange place. She bought a drop phone."

But Chapman made a crucial mistake. When she left the store a second time, she threw a Verizon bag into the garbage. The FBI team was watching her close enough to spot this. As soon as she left the area, a member of the Special Surveillance Group pulled the bag out of the garbage. Inside was a customer agreement for the purchase of a Motorola cell phone. It had been taken out in the name of Irine Kutsova. The address had been as fake as the name and was almost beyond parody: "99 Fake Street, Brooklyn." There was also packaging for two different prepaid calling cards that could be used to make international calls. The charging device for the phone was left in the bag. It had clearly been purchased as a burner for one-time use under a fake name. Chapman was worried. She no longer trusted her regular communications and wanted to make urgent contact. She had left the meeting and gone into panic mode. The FBI operation, at the last moment, was at risk of unraveling. Everyone was nervous but now this looked catastrophic. Was the illegals operation going to fall apart on the last day?

This was now a full-blown crisis for the FBI. "We believe that she was quickly processing that she may have come to the attention of the FBI," Todd A. Shelton of the bureau later recalled. "We were very concerned that if she sent a distress signal to the SVR they may in turn send an emergency signal to all of the illegals and we would lose them all." Senior FBI officials were looking at their watches and clocks. The arrests were due to take place across the country in less than twenty-four hours. But Anna Chapman had gone off the rails.

THE QUESTION FOR the FBI was what to do now. One option would be to bring everything forward by a day and arrest all the illegals. But there was a problem. Russian president Medvedev was still at the G20 in Canada. And the FBI was under strict orders from the White House not to move until he had left. They were straining at the leash. And there were real risks to the whole Ghost Stories investigation. If Chapman was suspicious enough there was still a chance she might contact Moscow Center and send an emergency signal, which, in the worst-case scenario, could lead to all the illegals grabbing a passport ready for just such an accusation and fleeing.

CHAPMAN'S MISTAKE WAS the receipt. It provided the FBI with the number of her phone. That night the FBI was up on the phone and able to listen as she made a series of dramatic calls on the way up to a boyfriend's place in Connecticut and through the night. And who did she call? Her father—the spy.

Something strange had happened, she told him, taking him

through the story. He listened carefully. So did the FBI. He sounded worried. That was bad.

He seemed more aware of the dangers than his daughter. At one point, when she revealed she had handed over her special laptop, he got angry. "You did what?" he lit into her. But what should she do? The first call ended. He tried to get hold of someone in Moscow—likely at the SVR—but it was the early hours of Sunday morning and senior officers were not around. When Chapman's father called back, his instructions were surprising. He recognized it might be bad, but she was not to panic. Instead she was to report the incident to the police. The FBI team would scratch their heads about his advice. Perhaps it was wishful thinking? For Anna, running now meant ending everything, giving up on life as a spy and the New York lifestyle that went with it. Rather than think that the operation was blown and it was all over, perhaps it was just easier to hope that this had been some prank by what he called a "hooligan" or perhaps at most a provocation. Even in the worst case, that she was under investigation, what proof might they have? Many years earlier, Kim Philby, the treacherous former MI6 officer, lectured a KGB audience and gave them his most valuable lesson after being confronted by his colleagues—never confess. They never had as much on you as you feared, he said. Her father, who perhaps had been in that audience, told an unsure Anna to bluff it out.

What would an innocent person do if a Russian had approached them and asked them to hand over a passport? They would report it to the police. That was what her father asked her to do. The problem was that this overlooked his daughter's mistake in handing over an operational laptop to someone she was going to claim now was a complete stranger.

There was still a risk for the FBI. They knew her father could call the SVR and ask if it was their person whom she met. They would say no. But the good news was that it was a weekend. It was now early Sunday morning in Moscow. If a message did get through to Yasenevo, it would only be to the duty officer. A more senior officer would need to be called on the weekend and would then need to make a decision. How serious did the situation look? How likely was it that she was really compromised? Her encounter was certainly odd, but it was hard to tell how bad it was. Pulling out one illegal after all the investment was not something to take lightly. And there was nothing to suggest compromises of any of the other illegals (whom Chapman did not know). Even if someone did think the worst, a decision to get her out was not going to be taken immediately. That was something the FBI had counted on. Doing the false flag the day before the planned arrests was deliberate. Just in case it triggered an avalanche of concern, they would still have time to move.

And of course, there was an additional factor in play. Who would the duty officer in the SVR call to discuss Chapman's concerns? One person they might call was the deputy head of the department that ran the American illegals. He was not going to be easy to get hold of.

In Belarus, Alexander Poteyev was on the move. He did most of the journey by himself and under his own steam. He now used a fake passport given to him by the CIA to travel, again by train, to western Ukraine, the Russians believe. The passport was in the name of a Viktor Dudochkin. There was a real Dudochkin who had handed his passport over to the US embassy in Moscow in 2009 for a visa application. The FSB would later believe that this had been secretly copied by the CIA. This would have been the most tense moment on the escape. If there were suspicions among the illegals that had been fed back to the SVR and someone had checked on his whereabouts, then they might be looking for him. Handing over a false passport to a border official also was a moment of jeopardy. But as each stage of the journey passed, Poteyev moved farther and farther away from the hands of his former colleagues.

It was a hot June Sunday, which did little to dampen the tension as the day of the arrests began. Much could still go wrong in the coming twelve hours before Medvedev's plane took off. The first question was whether Semenko and Chapman would fall into their traps and provide enough evidence to arrest them.

Just yards from where Semenko had been meeting, the whole operation was being coordinated out of the Strategic Information and Operations Center (SIOC) at FBI headquarters, on the fourth floor. This is where major events and unfolding crises are tracked. There are more than a thousand telephone lines as well as screens showing live feeds of surveillance and conference rooms for senior leaders to coordinate. Craig Fair was in day-to-day control. There had been twenty-four-hours-a-day physical surveillance on all the illegals for the days leading up to the arrest and especially once Poteyev was on the move. The nerves were jangling that a compromise of any of the surveillance by being spotted could also lead to an emergency signal. They knew every illegal had an emergency escape plan and fresh identities to support it. Anna Chapman would later claim that she had been suspicious that she was being followed in the days leading up to the arrests. She would claim her concerns were sent back to Moscow and to the man who was deputy head of Department 4. Poteyev, of course, would have ignored them.

There was also one additional fear. If the Russians had a penetration of the FBI and CIA they would have normally done everything possible to protect that mole and not act on their intelligence in a way that revealed their existence. But one thing they would not let happen is a large group of SVR officers getting arrested.

And so there was always this lurking fear that perhaps suddenly the illegals might all disappear as the net began to close, if there was someone on the inside who could tip them off.

A surveillance team was watching the dead-drop site in the park that Semenko was supposed to be servicing. They could not get close enough to observe in person so they had installed a camera. But there was no live feed, so the tension at the command center was palpable the next day as 11 a.m. approached. To everyone's relief, Semenko, in T-shirt and shorts, appeared and approached over a small wooden footbridge. He was carrying a white bag. At 11:06 he crawled beneath the bridge. It was awkward and would have looked suspicious to anyone watching. He kept coming out and glancing around. He removed the newspaper holding the envelope from the bag and placed it on the underside of the bridge. He looked around to make sure no one was watching and then left. Once they were sure he was gone, FBI agents went to the drop site and extracted the newspaper and envelope. The money was inside. Semenko had taken the bait.

Anna Chapman, though, was another story.

THAT SUNDAY AT 11 a.m., an FBI undercover officer was waiting near the World Financial Center in New York City. He was looking for a redhead with a magazine, ready to get confused about whether they had met in California or the Hamptons. The minutes passed. Chapman was a no-show.

She was never an early riser, even when fearing being exposed as a spy, and after staying at a boyfriend's place, she arrived back in Manhattan in midmorning. By lunchtime she had turned up at the 1st Precinct station of the NYPD in downtown Manhattan. "This really scary thing happened to me and I didn't know what to do," she told a police officer. It was a confusing tale involving a strange man, passports, and laptops. Because she had been scared, she explained, she had agreed with everything the man had told her to do.

Fortunately for the FBI, because they had been listening to the call with her father, they were waiting. Overnight and under huge pressure, they had cooked up a plan to stall Chapman. Two FBI agents had been sent to the 1st Precinct police station posing as detectives. One of them was now listening with great sympathy to her story. But they needed to play for time. It was still hours until Medvedev left North American airspace. It was a bizarre situation. They had a Russian spy in a police station and an arrest warrant in the room next door but they were not allowed to serve it because of orders from the White House. They had no choice but to play it long.

"Anna, what happened to you?"

She told the story to one officer. And she was asked about every detail. "Oh, my God, that is a crazy story. Let me get my partner in." In came the other officer. "Now, Anna, please tell the story again from the beginning." More questions. When that was over, it was on to phase two. "Here, look at these mug shot books. Tell me if you see anyone you recognize." They began to show Chapman photos of people she might have met in the cafe the previous day. She was obviously never going to spot the person, given that he was an undercover FBI officer and so were the two people she was talking to. But there were a lot of mug shots to go through. The whole charade went on for hours.

From Ukraine, Poteyev traveled to Frankfurt, Germany. From there he took a flight to the United States. He was back in the country that had recruited him all those years before in New York. The whole journey was monitored closely at CIA headquarters. Director Panetta was in touch with Robert Mueller at the FBI to make sure everything was coordinated. "Once we knew the source was safe we could proceed with the arrests," CIA director Panetta recalls (although like other officials involved in the operation he will not confirm the source's identity).

Finally, word came into the command posts on Sunday evening that Medvedev was in the air. The arrest teams were ready to move. In Chapman's case, they did not need to go far. They walked into the room at the police station and slapped on the handcuffs. Anna Chapman's spying career was over. But her time in the limelight was just beginning. And for the other illegals, after years, sometimes decades, living a lie, this was the day it would all end.

20

The Day It Ends

At the Heathfields' house in Boston, June 27 was a day of celebration. Their youngest son, Alex, had just come back from six months in Singapore and it was the twentieth birthday of Timothy, their eldest son. Donald Heathfield and Alex were also about to head to Russia via Paris, with Timothy following after. The family had been out to their favorite local Indian restaurant for a buffet lunch to celebrate. They returned to their suburban home on Trowbridge Street by 4 p.m. and opened a bottle of champagne. The brothers were tired from a house party the previous evening and were upstairs when there was a knock on the door. They were not expecting anyone, but their mother shouted up that it must be some friends of Timothy's who had come to surprise him. When she opened the door, there were people dressed in black. The brothers thought it might be a birthday prank. It took only seconds to realize it was not. Half a dozen men and women were at the door. "I remember vividly the FBI agents entering our house with weapons as I walked down the stairs," Alex Foley later said in an affidavit. "My parents were handcuffed in front of my eyes." FBI officers remember it differently. They say there was a knock on the door. Ann Foley opened it and stepped outside to talk to them. Everything was calm. If there had not been so many people there for the search, the neighbors might not have noticed anything amiss.

Heathfield and Foley said almost nothing. What the FBI team remembers about their reactions was that they were almost blank. It was almost as if they were physically there but mentally somewhere else. At the moment they came for him, Andrey Bezrukov had instantly understood his life as Donald Heathfield was over and two separate identities were collapsing into one: "It was as if all my previous life and plans had vanished in some kind of fog." Decades of patient work and sacrifice building a legend had evaporated in an instant. After nearly a quarter of a century, it was over.

Ann, now nearly Elena again, was taken out in handcuffs. She remembers the neighbors watching in astonishment. Who would look after her youngest son, she was asked. His brother, she answered. She and her husband were placed in separate cars and driven away. Timothy had been upstairs and one of the FBI officers walked him down. He was asked if he wanted to stay in the house while it was searched or go to a hotel room that had been booked. He and his brother opted for the hotel. When they came back to the house the next morning they found everything had been searched. Electronic devices and personal effects had been taken away—right down to Alex's PlayStation console. The family bank accounts were frozen. The children were left with about three hundred dollars each. Their teenage world had come tumbling down.

Their parents were taken to an office in Boston. It was Sunday and the place felt empty. "They kept us separate. We could not communicate in any way," Heathfield says. "When I came into the interrogation room, there was a huge box on the table. It had 'evidence' written on it and my name. Then I started getting nervous and frightened." This was the FBI's aim—to make clear how much evidence there was and that this was no speculative arrest. "I will not be able to talk my way out of this," he thought to himself. The question on his mind, like any arrested spy, was a simple but crucial one: why. "This question tormented us at first," recalled his wife, Elena.

"FBI agents treated me and my wife with studied respect, the way professionals treat other professionals." Heathfield always saw himself very much as the serious intelligence officer. He and his wife were confronted separately. It was explained what the charges were—espionage. She expressed surprise and denied it. "I am not talking to you about anything. I want a lawyer." They were facing twenty-five years in jail, maybe forty. But there was a deal on offer, he recalls: cooperate and it might be just five. It was a terrible dilemma on one level, as they knew their two children were out there in Boston. What were they thinking right now? If they faced the full force of the law, then their time as parents was over and the boys would be alone and bewildered. But there was no choice. They would not talk. They were taken to a solitary cell. There was a bench they could lie down on and a blanket. The husband and wife mentally prepared themselves for a new stage—one of arrest, detention, and endless questioning. This was one, they knew, that could last for years.

IN NEW JERSEY, FBI agents Maria Ricci and Derek Pieper were there in person for the closing scenes of the drama they had observed for so long. It was a steaming hot day—the worst kind for waiting for hours in your car for the signal that you can make an arrest. You wanted to put your head in a fridge just to try to keep cool. But after years of observing the lives of the Murphys in close focus—listening to their conversations and walking around their home when they were away—they were not going to miss the chance to finally meet face-to-face.

The Murphys had just settled down for dinner when two FBI

teams arrived at the house. Richard Murphy opened the door and saw a sea of blue "arrest" jackets with the distinctive FBI logo on them. He instantly understood. There was no shock. Today was the day it all ended.

"A lot of times when we arrest people, it is like 'I didn't do it.' When they came to the door they were very calm. They saw all of us. And it was this look, kind of like 'well I guess today is the day. This is happening,' " says Ricci.

Murphy even tried to crack a joke. He had a drink in his hand. "I just opened my beer," he said, joking to see if that might give him a few more minutes.

As the time for the arrests approached, Ricci and Pieper had been debating over who would interview each of the Murphys. Who would be most likely to build up the best rapport and get them to talk? They settled on Ricci with Cynthia, but at the last moment they switched to Pieper. What they were unsure of was whether she was going to play the KGB colonel or the mom who was concerned about her kids.

When they were drawing up the arrest plans, the FBI had realized that there was one job description they needed but did not have among their agents—babysitter. How to deal with the kids had been a major concern for the FBI. In the case of the Murphys, the view was that it would not be acceptable for them to be taken away to a child welfare center. The agents, like Ricci, had watched the Murphy girls grow up. They did not want the kids to be sent to some child protection agency. The girls were innocents and the whole experience would be bewildering and bad enough for them anyway without that. Taking proper care of them was about showing a touch of humanity even for your adversary.

Eleven-year-old Katie Murphy was away at a pool party for a friend's birthday and the first few minutes of the conversation inside the house were about the children. Pieper tried to reassure Cynthia that the children would be okay. She was asked if there was a friend who could look after them rather than being sent to children's services. In another sign of how well the FBI team understood them, they already had a pretty good idea of whom she might suggest and were proved right. The hope was this would relax her. But she remained cold. With the initial conversation over, they headed outside of the house to the cars. Cynthia said little. As they went past the garage door it was open. It needed an electronic code to close it. Cynthia asked Derek Pieper if he needed the code to put it down. "We have that," he said. Her face was a picture. The Russian spy now realized the FBI owned her life—down to the code for her garage door.

Meanwhile, Ricci was with Richard Murphy in the house. For all his gruffness on the phone and his snippy emails to Moscow Center, he ended up being personable and warm. There was some small talk, but he was clearly deeply worried about the kids and seemed more concerned than Cynthia about what was going to happen to them. He kept telling Ricci to make sure Katie took her medicine, which was in the fridge. "While Cindy was the intelligence officer, he switched quickly to dad," recalls Ricci. Since dinner had been interrupted, Ricci said they would get some food when they got to New York. "Can I get a beer, too?" he joked. "I don't know if I can get that. But I'll do my best," replied Ricci. The pair were driven away separately.

Katie was still in her bathing suit and clutching an animal floater when she came home on Sunday evening. She found her sister Lisa still in the house but her parents gone. Sitting with her sister was a female FBI agent, while others were combing through her house. The two daughters were later taken away in a minivan with tinted windows, carrying pillows and rucksacks. It must have been bewildering for girls so young, who would have not even the slightest conception that their parents were anything other than what they had seemed to be. But now suddenly, they were being ripped out of the quiet, suburban lives they had known. The girls looked stunned, neighbors remember.

The neighbors stood outside the Murphys' home in shock after the arrest. Journalists would find them debating whether or not to water the garden. Some questioned whether it was wrong to water the hydrangeas if they belonged to Russian spies. "The hydrangeas did nothing wrong," responded one. The spies were gone and the suburbs closed up after them as if they had never been there.

CYNTHIA WAS DRIVEN from Montclair through the Lincoln Tunnel and into Manhattan. The drive—even on a Sunday evening—was long. There was a pride parade in New York that weekend, which added to the journey and the slightly surreal element. Pieper still tried to focus on her role as a mother in an attempt to relax her—is there anything else the kids need to bring along with them? All he got in reply were one-word answers.

In many spy cases, gaining a confession is vital since the suspect has often been identified based on secret intelligence—like a CIA source or an NSA intercept—that cannot be revealed in court. But with the illegals, that was not a problem, since the FBI had been building a case for years. But still, out of sheer human curiosity rather than legal necessity, the FBI officers involved in the case were hoping that the illegals might elect to talk.

When they got to the office, Cynthia Murphy was fingerprinted. Pieper's attempts to charm her had met with a blank. Now it was time to confront her with the full extent of what the FBI knew. A stack of photos was laid out in front of her. They were a gallery of her life over the last ten years—a clear signal that the FBI had been watching all along. She remained unmoved.

"I know what your job is," she said to Pieper. "And I don't want to talk to you." That was that. *You know what, I'm not going to beg,* thought Pieper. *That's fine. Enjoy the rest of your night.*

There was a human curiosity about what these people were really like and also a professional one—as fellow intelligence officers what had it been like? The FBI team wondered if they could just forget the handcuffs for a moment and say, "It's over—let's talk." But it never happened. They had expected that Sunday evening had just been an initial interview and that there would be the chance for more interviews in the coming weeks. So they had held back, hoping to work them slowly. What the FBI team did not realize was that there would be no second chance to try to persuade the Murphys and others to talk. Events would move faster than anyone expected.

SEMENKO WAS ARRESTED at his apartment in Arlington. A neighbor saw him being taken away by FBI agents with his T-shirt pulled up over his head. The third true-name illegal, Alexey Karetnikov, was picked up by Immigration and Customs Enforcement (ICE) agents on the twenty-eighth. He had not been in the United States long enough for the FBI to be able to charge him with a crime. The best option, it was decided, was the quieter way of dealing with spies: the FBI had a simple word with him to let him know he was busted. He was formally deported to Russia after admitting he was in violation of immigration law, and it was made clear he would be arrested if he returned without permission.

Zottoli and Mills were detained at their Arlington apartment. There was only one person who knew their children. The problem was that she happened to be flying to Washington, DC, for a conference at the time of the arrests. So the FBI put an agent on her plane who could turn to her at the right moment, explain what was going on, and ask if she could look after a couple of children. Mills, some of the FBI agents remember, was the tougher of the couple when it came to the moment of confrontation.

IN YONKERS, VICKY Pelaez and Juan Lazaro had been out at a barbecue in New Jersey on Sunday and an FBI car intercepted them in front of their house as they were returning. Pelaez would later describe it in the most dramatic of terms. "Dozens and dozens of heavily armed FBI men, dressed in black, many masked, burst into my life in what seemed like a movie," she wrote. When their two children—Waldo Marsical and Juan Lazaro Jr.—made it to the house they were in for a shock. The FBI was everywhere. The agents wanted to know if there were secret hiding places. The children were astonished at what the agents already knew. "They knew my nickname," Juan Lazaro Jr. told his mother's newspaper. "They knew absolutely everything." Even, he said, that he was considering applying to music school, adding, "They asked me if I'd looked at conservatories in Russia."

For some elderly people, news of the arrests would only bring pain. As soon as they could, FBI agents traveled to Canada to see the real parents of Donald Heathfield and Ann Foley to explain to them that their private grief was about to become a detail in a spy story. Old wounds, never fully healed, were reopened. There were other hurried conversations as well. Alan Patricof wanted to give Hillary Clinton at the State Department a heads-up, knowing that her name might emerge when it came to his contact with Cynthia Murphy. Clinton's office would issue a statement saying there was "no reason" to think the secretary was a target of this spy ring.

THERE WERE OTHER moving parts in play in an intense few days. Russian sources say that on Sunday, June 27, a female FBI agent called an SVR officer who was based in Chile. The phone number they had used was known to only two people—the officer's wife and Alexander Poteyev. The FBI will not comment but Russian sources say the officer was offered $100,000 to turn. Instead he fled as quickly as possible through Argentina and back to Moscow. He had been another one of those identified thanks to intelligence from their source.

And of course, as the news of the phone call and arrests filtered back to Moscow, there would have been anguish at Yasenevo, with the realization that decades of work and investment had been undone in a single day. It is hard to imagine what the scene might have looked like as officers arrived at work in Directorate S on Monday morning to the news. An urgent meeting would have been called. But why could no one get hold of the deputy head of Department 4?

ON JUNE 29, Christopher Metsos checked out early from his cheap hotel and headed to Larnaca airport. There, according to the FBI plan, he was stopped by Cypriot police on a US extradition warrant as he was about to catch a flight to Budapest, Hungary. His female companion continued to board the flight. But then things began to go wrong. Metsos was taken to court. He seemed strangely calm and showed no anxiety, the lawyer who was assigned to him recalled. "He told me that he had nothing to do with this case. He didn't understand why he was there." The pair had coffee and water as they waited. In a twenty-five-minute hearing, the judge agreed to let Metsos out on $34,000 bail until an extradition hearing a month later. The money was quickly handed over. He was supposed to hand in his passport as well as turn up at a police station every day. This was a man who had a dozen identities and passports to go with each, as well as the backing of the SVR and Directorate S's resources and experience. The idea that handing in a single passport was going to keep him on the island was absurd. He registered at the local police station and calmly went back to a new hotel, the Achilleos, at around two thirty in the afternoon. He paid 630 euros for a two-week stay. He hung a "Do Not Disturb" sign on the doorknob and took a shower.

The next day he failed to report to the police between 6 p.m. and 8 p.m. as instructed. It was only the day after that the police entered his room. They found the bed still made and a pair of his slippers sitting next to it. None of the staff had seen him leave. Perhaps he had slipped away while the night porter was in the bathroom or perhaps he had climbed over a balcony. However he did it, he had done it quickly and with no one seeing a thing. A proper spy's escape.

The United States was not happy. "As we had feared, having been given unnecessarily the chance to flee [Metsos] did so," said Dean Boyd, a spokesman for the Department of Justice. The Cyprus authorities went through the motions of trying their best to look for him. They said all exit points were monitored. A police photo that showed him looking impassive and sunburnt was distributed. But no one had the slightest expectation that he would be located.

The best guess is that he headed to the northern part of the divided island and from there by boat to Turkey and perhaps Syria before going to Moscow. Others wonder if Russian organized crime groups involved in human trafficking got him to Greece, where the SVR could have whisked him away. Cypriot government officials insisted that the decision to give bail to Metsos had been one for the courts alone. But many in the United States believed this was a favor from the Cypriot government to the Russians. As the search was under way, the country's president was hosting a reception celebrating the arrival of a Russian bank on the island. President Demetris Christofias had been educated in the Soviet Union, met his wife there, spoke Russian, and was a self-professed friend of Moscow. He was looking forward to a visit from the Russian president in October of that year. Even though the United States and Britain use Cyprus as a spy base for operations into the Middle East (with a large British listening station), Russian influence runs deep. It was one of the places Russians stashed their money and Limassol was known as "little Moscow." In 2013, when the Cypriot banking system was close to collapse, rather than take an EU bailout, which would have carried with it demands to clean up the system, the country turned to Moscow.

The FBI team was not bothered as to whether the illegals they arrested would talk or not. After all, they knew pretty much everything about their lives already. But Metsos was different. If he had talked they may have been able to learn much more about illegals around the world. He would have been the ultimate prize. "It would have been huge to talk to him," says Alan Kohler. But Metsos, ever the wily operator and veteran spy, slipped from their grasp. There remains deep anger about that among those who worked on the case. And he has never been seen since. Like a ghost, he simply vanished. That only left those languishing in prison cells in America. "In terms of value to the FBI in preventing this happening again, there wasn't a lot there," says Kohler. "Their value was in a trade." And now the question was whether that deal could be done.

21

The Squeeze

"WE'VE GOT THEM by the balls," Mike Sulick said. "Now let's squeeze." Sulick, a veteran of Russia House, had risen to become head of the National Clandestine Service of the agency and was talking to his boss, CIA director Leon Panetta. The veterans of the spy wars with Russia had the upper hand with their old adversary and they wanted to stiffen the sinews of those around them. As Panetta picked up the phone in his Langley office two days after the arrests, he was going to hear what the squeeze sounded like. On the other end of the line was Mikhail Fradkov, the head of the SVR. Panetta, who liked the odd spy film himself, thought his opposite number was straight out of central casting. There had been some surprise when Putin had made Fradkov prime minister in 2004 and then head of the SVR in 2007. That had made some wonder whether his years as a trade and economic official back in the 1970s and 1980s had been cover for a KGB career.

Panetta and Fradkov had been trying to build some kind of working relationship. It was the old story of the attempt at dialogue and liaison. But the ghosts of the past were always there. Fradkov had been Panetta's guest at Langley a few months before the phone call. The two looked at the photographs that line the walls outside the CIA director's office. One was a picture of

Oleg Penkovsky, the GRU colonel who had provided secrets to CIA and MI6. Fradkov visibly winced. That night they went to a restaurant. The conversation was stilted. There was vague talk of cooperation but no common ground on what that meant. Old Russia hands in the CIA, who remembered the misfires of the 1990s, were deeply skeptical about discussions of creating a joint working group. "The old Cold Warriors in the CIA's clandestine service—including Mike Sulick—told me I was wasting my time," the CIA director later reflected. They said the Russians would never share anything and just use it as a ruse to try to recruit the officers involved in liaison. "These are a hard-knock group of people who have been doing this for years," Panetta said of his Russia House staff. He knew they had never bought the idea that the Russians had changed their tune with the end of the Cold War.

What did Panetta think had been his country's worst intelligence or foreign policy mistake, Fradkov asked Panetta over dinner? The mismanagement of the Iraq War was the American's answer. Panetta asked the same question about the Soviet Union. Fradkov paused for a long moment, and then answered with one word: "Penkovsky." The greatest mistake was not a policy choice like the Soviet Union invading Afghanistan, but someone within the intelligence services who had betrayed. That said a lot about how treachery was viewed in Moscow.

Panetta, accompanied by Sulick and a few others, had also been over to Moscow. "I think I can still hear the screams from the basement," his chief of staff whispered to Panetta as they walked through the Lubyanka. During that visit Panetta carried a secret. Just weeks before, the FBI had briefed him on the illegals they had under surveillance with the news that the operation was moving toward the arrest phase. While touring Yasenevo and meeting Fradkov, Panetta and Sulick had kept quiet that they were about to deliver a crushing blow to the man they were meeting.

Now on the phone, Panetta was going to broach the idea of a swap. There were many obstacles. But one of the biggest was that the Russians had to admit that the people the FBI had arrested were their spies. The SVR would have to blow their own carefully constructed cover. Everyone knew this would be a difficult moment. The capture was a disaster when Russia's spies were already on the back foot amid concerns that President Medvedev was going to reduce their power.

The Russians' initial response to the arrests had been predictable—the charges were "baseless and improper." But one reason the indictments were so detailed and broad was precisely to confront Russia with the truth—the Americans had not just caught their illegals but knew everything about them. By revealing that some of the surveillance went back a decade and included coverage of their houses and clandestine meetings, the Russians would hopefully be forced to face up to the reality that the illegals' cover was blown.

Panetta used the speakerphone for the call. Around him were the top Russia hands from the CIA. Expectations were low. "These are old Cold War warriors who had been through a lot," Panetta recalls. The assumption was that the Russians would deny everything.

"Mikhail, we have arrested a number of people, as you saw in the press. Those people are yours," Panetta said. A CIA interpreter translated the words. Then there was a long pause on the other side. If the Russians did not acknowledge the illegals, there could be no swap. There was a long pause before Fradkov replied.

"Yes, they are my people."

"The men and women around me had to stifle themselves to keep from cheering out loud," Panetta later said. "Instead, a silent round of raised eyebrows and high-fives ran through our room as we realized that we were already past denial and into negotiations." Panetta made his offer: "We're going to prosecute them. If we have to go through trials, it is going to be very embarrassing for you." There was an element of bluff in this since the White House had indicated they did not want high-profile prosecutions damaging the relationship.

There was another long pause.

"What do you have in mind?" Fradkov asked. The Russian sounded humble on the other end, one of those involved recalls. It was clear he had been given his orders—get the people out.

"I propose a trade," Panetta said.

Fradkov did not immediately agree. He said he would check. "From our side, we believed he had to check with Putin," says Panetta. The game was on.

The SVR was in a tough corner. The whole point of illegals was that they were "illegal"—they had no diplomatic protection and cover. Once caught, the option of maintaining their cover by not admitting the truth risked consigning them to a trial and—given the scale of the evidence the FBI had already presented—an almost certain prison sentence. This was where the culture of Russian espionage again came into play. Illegals were heroes. They were not agents, people like Hanssen and Ames who had sold out their country for money—they were officers. They were patriotic Russians. To leave a hero rotting in an American jail was not a good option. The CIA knew this. That was why the old Russia House hands like Mike Sulick wanted to squeeze.

The United States had the upper hand. Word soon came back from Fradkov that a deal would be done. Normally this took time, but no one wanted to wait. The Americans were surprised at how receptive Fradkov had been to a swap. So how could the CIA use its leverage?

They decided to ask for something unprecedented. There had been plenty of spy swaps before. But those in the Cold War were typically a Russian caught in America exchanged for a Westerner caught in Russia—like British businessman Greville Wynne, who had been involved in the Penkovsky operation, who was traded for KGB illegal Konon Molody (also known as Gordon Lonsdale). Or downed American U-2 pilot Gary Powers for the illegal Rudolf Abel. But what was being proposed in 2010 was going to be different. The Americans wanted Russians to be swapped for Russians. Moscow was going to be asked to give up its own citizens—people it regarded as traitors. This was much more painful for Moscow and its intelligence services. It was the equivalent of Russia asking the FBI to free Robert Hanssen from his prison sentence.

The priority for the CIA was Alexander Zaporozhsky. He had been an important asset for the CIA and the agency had long wanted him back (even though a few inside were annoyed that he had got himself into trouble by returning to Moscow). The Russians knew how bad the Americans wanted him. They had, it can be revealed, offered to swap him for a number of years. But the person the Russians wanted in return was someone the CIA would not give up—the agency's very own traitor: Aldrich Ames. The Russians had raised this year after year. And year after year they had been rebuffed. Now the CIA had the chance to get Zaporozhsky out without giving up their own traitor. But one was not enough. Everyone knew that would look like a bad deal. "We had ten people plus their families and we thought if this thing was going to be able to be justified, it had to be a group of people on our side that were going to be exchanged," Panetta says. "This was a good opportunity to get as much as we could."

A second name was added—Gennady Vasilenko. He had played an important—even if unwitting—role in helping catch Robert Hanssen by helping get the man who sold the file over to the United States. There was a sense of guilt for the fact that Zhomov had made him pay a price for that. But that was only two.

It was time to offer a favor to some friends. Mike Sulick called his opposite number, the director of operations at MI6, who had worked the Russia target back in the 1990s. Sulick asked if there was anyone MI6 wanted out of Moscow. The MI6 man went to see John Sawers, the head of MI6, in his sixth-floor office at Vauxhall Cross. Sawers was six months into his job as "C" or chief.

Britain had not been told about the plan for the arrest of the illegals. But the news was met with undisguised pleasure in London when it broke. MI6 had no direct relationship with the SVR—post-Litvinenko it had wanted a back channel to send messages, but MI5 had quashed the idea. Cooperation between the CIA's Russia House and their counterparts at MI6 was close, aided by the fact that both sides knew the other had to keep secrets even from their own colleagues to succeed. Sawers's people came up with two names. One was Sergei Skripal—he was their agent and they felt they owed him a debt. The other was Igor Sutyagin. In his case, there was a sense of guilt inside Vauxhall Cross. He had paid a heavy price even though he had always denied working for them. Sutyagin's name had also come up in Washington in another context. Mike McFaul, the senior director for Russia at the National Security Council, had known the Russian when he visited Stanford in the early nineties and had followed his case.

So now it was four. Even that, some reflect, was not as good a trade as it could have been. "We didn't get enough," says one person involved at the time. But then perhaps it was a good thing there were not more people festering in Russian jails having been convicted of espionage for the United States or United Kingdom.

MI6 and the CIA may have had a unique opportunity to get people out. But delivering a deal was not going to be straightforward. The problem was one that few would have predicted. Those accused of crimes had to admit the truth for the deal to work. And on both sides there were some people who did not want to be swapped. They did not want either to plead guilty or to go where they were told.

The illegals, held in Boston, New York, and Washington, had been read their Miranda rights on arrest. According to material provided to the court, Zottoli and Mills waived their rights and admitted these were not their real names and that they were Russians. But they did not say who they really were. And others were also reluctant to own up.

On the night retired spy Lazaro was arrested, he talked for several hours to the FBI and made a lengthy statement. He was not born in Uruguay, nor was he Juan Lazaro. He said that his wife had indeed gone to South America and delivered letters to the "Service" on his behalf. The house in Yonkers had been paid for by the "Service." He said one more surprising and personal thing. He said that although he loved his son, he could not violate his loyalty to the "Service" even for him. He was a professional doing his job. But there was one other thing he would not do. He refused to give his real name. FBI agents thought the experience of talking was almost cathartic—a release after so long living a lie. A bed was made for him in the office but when he woke up the next morning, he stopped talking.

At a July 1 hearing, the government made the case for the illegals not to be given bail. The fear was that the SVR could whisk them away (an argument helped by the fact that Metsos had fled). Since the illegals did not have access to classified information and were not being charged with espionage, the criminal complaint had to be carefully drafted. Since they had used their true names, Semenko and Chapman were charged with failing to register as agents of a foreign government—with a possible sentence of five years. The family illegals were charged with acting as unregistered foreign agents and it was alleged they had used false identities to illegally move money to obtain mortgages and rent properties, allowing an additional money-laundering charge. The white-collar fraud charges offered a potential for significant jail time—up to twenty-five years. That, in turn, would raise the pressure on Moscow to come to a deal.

Elena Vavilova was kept in solitary. The air-conditioning was on full. "It was cold and lonely," she later said. There was plenty of time to pace the cell, to speculate about what had gone wrong and worry about how many years she might be facing behind bars. But there was no tough interrogation. And she soon understood why. As they read the criminal complaint against them, the illegals realized that this had not been a lucky break by the Americans. The scale of the evidence was stunning. "When I read the paper and saw lots of people's names and place names there, it became obvious that one single mistake made by one person could not have brought about this catastrophe," Vavilova later said. On the one hand there was relief that they had not been caught because they made a mistake. But on the other, there was the realization that it had been due to something much worse. They had been betrayed. There was a mole.

This was a difficult time for the children. One minute they were normal American kids, next they were the sons and daughters of Russian spies. The entire worlds that had been constructed around them suddenly collapsed amid a flurry of tabloid headlines and breathless news reports. At the hearing, Vicky Pelaez cried and gestured to her two sons from the courtroom. Her children maintained the whole thing had to be some kind of mistake as journalists scrambled to find out what they knew. "It's a circus. This is pure psychological pressure. It's total confusion. He's an old guy. His English isn't so good," Waldo Marsical said of his stepfather, Juan Lazaro. The charges were "ridiculous," he said, explaining that his parents were clueless when it came to computers and technology and often needed help accessing their Yahoo email accounts. So how could they manage clandestine communications with Moscow?

At the first hearing, the two sons of Donald Heathfield and Ann Foley had watched in shock as their parents entered shackled and wearing orange uniforms. "I never had any doubt about my parents' identities," Alex later said, before adding something that not many teenagers could say about their parents: "Never once in my 16 years of life had they done anything that seemed odd, or unexplainable." At the hearing, the boys had managed to talk briefly to their parents. They spoke in French so that people around them could not understand. "We told them to try and leave the city," Vavilova recalled. She says she found it hard as a mother to see her children look so scared. But of course it was her choice that had put them there.

On the fourth floor of the New York courthouse were a US Marshals Service office on one side and a holding area for defendants on the other, with a corridor and some seats in between. As lawyers for the illegals arrived, they noticed two men in the corridor in suits pacing and sometimes talking in hushed tones. They were there when the lawyers arrived and there when they left. In a highly unusual move, the two men also went into the holding area to talk directly to the illegals without their lawyers present. The men, it emerged, were from the "Russian Government." They looked like Russians, the American lawyers thought. They were there to cajole and pressure their people to sign up for the deal. It was not going to be easy.

Heathfield and Foley initially had brushed away all the accusations against them. They were who they said they were and nothing else. A fellow inmate approached Heathfield in prison after a couple of days and showed him a newspaper story about Russian spies with his photograph. "Is that you?" "I said it was some mistake," he later recalled. That all changed when they were paid a visit by a Russian official. He explained that a deal was in the works. They had permission to blow their own cover and admit their real names. After nearly a quarter of a century undercover, it was over.

Lazaro and Pelaez would prove the real problem. Vicky Pelaez was not a Russian. She had never lived in Russia. She had not even, she said, known her husband was Russian. Her lawyer said he believed she would not want to go to Russia. Robert Krakow, meanwhile, had been assigned as a lawyer to Juan Lazaro. He found him smart, intelligent, and engaging. They talked a little about Lazaro's university work. Lazaro seemed surprisingly calm at the way events had turned out. He also did not seem to want to go to Russia. He had two grown-up kids and a life in America. And he was retired from spy work. "What am I going to do in Russia?" he asked.

The illegals were told they did not have a choice. There was a deal and they would all have to accept it. For Heathfield and Foley and others this was straightforward. They were going to follow orders. Others were less keen. Lazaro was pressured by a heavy-handed Russian, his lawyer said at the time. "His manner was: 'This is what's going to happen.' " Pelaez was promised she would get two thousand dollars a month and housing if she went to Russia, although she would be free to go elsewhere if she wanted.

Lazaro and Pelaez had a peculiarly intimate problem. She has always maintained she did not know her husband was a Russian illegal whose real name was Mikhail Vasenkov. At one point, she spoke direct to her husband in custody. "What's your name? Your real name," she asked him, according to lawyers who were present. "My name is Juan Lazaro," he replied. The fiction, it would seem, remained. And while it was there, the swap could not take place.

. . .

IN MOSCOW THE CIA's chief of station, Dan Hoffman, was having his own problems. Panetta and Fradkov had agreed to the overall deal but had left their declared officers in the other's capital to work out the details. For the CIA, they knew closing the deal was important. If something went wrong and the deal collapsed, the four imprisoned Russians they had identified as wanting to leave could be in real danger.

Hoffman, a tough character who did not shy away from a fight, found himself sitting down to negotiate with someone the CIA knew all too well. There across the table was Russia House's long-standing nemesis—Alexander "Sasha" Zhomov. More than two decades before, he had been PROLOGUE—the young KGB officer who tricked the CIA as a "dangle," pretending to spy for them in Moscow while feeding false information. Later, as the KGB and then FSB's chief hunter of American spies, he had arrested their agents. And Zhomov would have known much about Hoffman—down to the most personal details. The American had served an earlier tour in Moscow as a young officer and so detailed surveillance reports about his life would have been delivered to Zhomov. Now these two adversaries had to work together to deliver a deal.

For Zhomov the arrest of the illegals stirred conflicting emotions. On the one hand the arrests were a disaster at the hand of an adversary he had spent his life battling. They had the upper hand now. And they wanted to free Russians whom Zhomov had dedicated himself to capturing. In the case of Zaporozhsky and to a lesser extent Vasilenko, these were men he held responsible for betraying some of Russia's best agents. It had been Zhomov who had raised with the Americans for years the idea of swapping Zaporozhsky for Ames. It may well have been Zhomov who came up with the idea of luring the former KGB man back to Moscow with just such a plan in mind. Now he was being forced into a trade, but not on the terms he had envisaged. However, spying is a complicated business. And there was an element of schadenfreude—taking pleasure in someone else's misfortune—for Zhomov. After all, Zhomov was FSB. And the illegals were SVR. Their roll-up offered a chance for him and the FSB to press their case that the SVR could not be trusted to look after its own internal security as it had done in the past. In the bureaucratic battles that marked out Moscow's internal spy wars, this defeat for a rival was also his opportunity. The wheels began to turn.

ON JULY 5, Igor Sutyagin was hard at work in a penal colony in the northern region of Arkhangelsk, near the Arctic Circle. Some days he was winding wire around a cable, other days hauling wood. That day he was shoveling cinders onto a path to make a walkway. An official came up to him. "Get your things together quickly," the official said. "You're being sent away." He was not told where or why. He accounted for his possessions, which consisted of 23 spare buttons, 17 paper clips, 106 postcards, and 74 envelopes. Sutyagin was handcuffed and put on a flight to Moscow, where he was transferred to Lefortovo prison. Being moved around without being told was a normal part of the punishment regime. But when he was asked what size shirt he wore, he sensed something unusual was going on. A prison official even did up his tie for him. He was aware he had not shaved for days, though, as they took his picture. It was the kind of picture you take for a passport or a visa. *What are you bastards up to?* he thought.

GENNADY VASILENKO HAD been in a prison in Nizhny Novgorod when he was given the sudden order on July 5 to pack up for a transfer. He was placed into a steel cage in the back of a VW. The smell of fresh paint and heat on the drive to Moscow was overpowering. By the early hours of the sixth he was back in Lefortovo. Skripal and Zaporozhsky also made their way to Moscow.

Sutyagin was taken to the office of the head of the prison. Inside were some Americans and Russians. They used only their first names and Sutyagin did not know who they all were, but one was Hoffman, another Zhomov. Sutyagin was told his name was on a list submitted by US officials and there was the possibility of his release and transfer out of the country. But in order for him to be pardoned, he had to sign a confession. He refused. He did not want to admit to being a spy. He also did not want to leave Russia. His dream was to be a free man in his own homeland. Both the Russians and Americans explained to him that he would have to sign if he was going to get out. "It's a very simple deal: you give your honor in exchange for your freedom," he would reflect. He could sense from the men in the room that larger forces were at work. He was told the exchange had been agreed to at the highest level—by the presidents of Russia and the United States and then prepared by their secret services. He was told if he said no, then the whole deal was off. He was given two hours to think about it. The pressure was on.

That morning Vasilenko was taken to the same room. He recognized Zhomov, who had interrogated him. One of the Americans introduced himself as the chief of station for the CIA in Moscow. This seemed bizarre. Why would a CIA officer be identifying himself in front of the FSB? Was it a Russian provocation to extract a confession that he really had been a CIA agent? Hoffman read out the details of the agreement between the two governments. If he signed, Vasilenko would be offered a pardon. He was told there was no time. He did not want to admit to espionage, as he had never been a spy. He said he needed time to think. Skripal was brought to the upstairs office. He did not want to confess his guilt, either. But he was told he had to. Hoffman told him, too, to go away and think about it. What about Zaporozhsky? He is likely to have been the one who was happy to sign and get out of Russia as soon as he could.

After a few hours Vasilenko decided he had nothing to lose. Skripal had come to the same conclusion and so eventually, reluctantly, would Sutyagin. "I was between a rock and a hard place and if I didn't sign, the rock and the hard place would have pulverized me," he said afterward. Some Russian officials, including the head of the FSB, were said to have been delighted by his confession. The decision to sign something admitting guilt for something he felt he had not done left Sutyagin depressed and angry.

Hoffman finally had all four signed up. The agreement was that they would be pardoned and allowed to leave with one member of their family. The last twenty-four hours were bad for Zaporozhsky. He was beaten and, according to one source, subject to mock execution. Vasilenko was also mistreated. "They really took it out on them," says one American. It was a sign of just how much anger there was in Moscow at being forced into a trade releasing men they saw as traitors.

On the eighth, the prisoners were expecting to be freed. They were taken to a reception room and allowed to change clothes. The hours passed. No one came for them. Only in the evening were they told that they would have to wait another day. They changed back into prison uniform and returned to their cells. Was it all a trick?

IN THE UNITED States, all the illegals had been transferred to New York. Many of the ten had never met each other before. There were no embraces and little emotion. They spoke in English rather than their native Russian. "It is difficult to explain, but you get so accustomed, you live that life," Elena Vavilova later said. Lawyers were stunned at how fast events were moving. The illegals were told they were heading for Moscow and their children would be joining them. Some had been told who their parents really were, and did not want to go to Russia. But what if some had been recruited to follow in their footsteps? The decision was to send them all to Russia with the exception of Pelaez's children, since she had not been an illegal and they were older. Juan Lazaro Jr. struggled, it was said, when he had to say good-bye to his parents. He was a teenager with a musical scholarship to his name and a glittering career ahead of him who suddenly was in the middle of a spy drama. When reporters outside court asked if he was going to follow his parents to Moscow he managed only a half smile before heading down to the subway. His half brother did his best to protect and support him. As part of the deal they had to pack up and leave the family house (paid for by the SVR). But they still could not quite take it all in. "I still believe Juan Lazaro is from Uruguay," Waldo Marsical told reporters the day after his father revealed who he really was. "The only Russian thing my mother likes is vodka with passion fruit."

ALEX AND TIMOTHY, Heathfield and Foley's two sons, were already in Moscow. The two young men were bewildered by the speed of events and the sudden realization that their parents were not who they thought they were. The brothers traveled to Russia on July 5 after their mother told them to continue with their planned trip. "If I wasn't already feeling traumatized enough by the events that were taking place, arriving to Russia truly made it all feel unreal," Alex later said. As they walked off the plane and before even going through passport control, they were met by SVR officers. The officers explained that they knew their parents and were going to help. The Russians asked the boys to trust them. Instead of going to the hotel they had booked, the brothers were driven to an apartment where they would stay for the coming weeks. "They showed us pictures of our parents and of their childhood," Alex later testified. The photos included pictures of their parents looking younger and in uniform and wearing medals. "That was the moment when I thought, 'Okay, this is real.' Until that moment, I'd refused to believe any of it was true," Alex said. His parents were not Canadians but Russian spies. This was their parents' homeland, a country in which the two brothers had never set foot. The boys were taken to see their grandmother—the mother of Donald Heathfield, or to put it more correctly, Andrey Bezrukov. She was a woman they had no recollection of meeting before (although they may have met once when they were babies). She spoke no English and they did not speak Russian.

IN NEW YORK on July 8, the illegals had to plead guilty. They would have to promise never to reenter the United States without agreement from the attorney general and to not profit directly or indirectly from the stories of their time in the United States. Their property would be forfeited. The Murphys had to give up the Montclair house, which they had valued so much it had led to a row with the SVR. But most important, they had to admit who they really were.

As sentencing approached, Juan Lazaro was still holding out. But the pressure was growing to admit the truth. That was required so that a guilty plea could be entered and the swap carried out. Finally, the barriers came down. Just before the sentencing, he finally relented. Pelaez maintains she was shocked to learn that the man she had been married to for a quarter of a century—the father of her child—was a KGB-trained Russian spy. "I had mixed emotions in that cell, where nobody told me what was happening. I am a nearly 60-year-old woman. I love my companion. But I may never forgive him for not being straight with me," Pelaez added.

The courtroom was packed on July 8. Two Russian officials

were present to make sure everything went according to plan. For Maria Ricci and others in the FBI team, the appearance in court was one they would never forget—the culmination of years of hard work. There were mixed feelings in the FBI about the decision to swap rather than prosecute. There were certainly some who wanted a guilty verdict and a prison sentence. That was what normally constituted success in the bureau. But there were those who knew this was different. Some admitted a sense of relief they would not have to go through the paperwork of the trial. And bigger political forces were at work. Sometimes success was not in the sentence, one agent involved in the arrests reckoned, but in the fact that your case makes it all the way up to the White House and results in the president deciding how he is going to use the leverage you have created.

THE SUSPECTS HAD to stand up in court and formally plead. Some, like Patricia Mills, looked broken. Others seemed to enjoy the attention. "As I moved to New York, I failed to register as an agent of a foreign government," said Anna Chapman. Reporters in the court recall her "breaking into a smug grin at times and twirling her red hair as she turned to look at the courtroom sketch artist." Her lawyer afterward said she was expecting to go to Moscow and then return to London. Perhaps her Western party lifestyle was not quite over, she was thinking. She would be disappointed.

Richard Murphy had been first. He stood up. The judge asked him to state his name for the record. Murphy looked at the judge. "Which name?" he asked. "That is one of those moments that will stick with me forever," says Maria Ricci.

22

Vienna

At 4 a.m. on July 9, the activity began at Lefortovo in Moscow. The prisoners were told to get their bags. Two hours later, they were taken to a van. "We've got to stop meeting like this," Zaporozhsky said to his old colleague Gennady Vasilenko when they sat down inside. Under heavy guard, the convoy made it out toward Domodedovo airport. Their van drove directly to the airfield where the Yak-42 Emergency Ministry plane was waiting.

The Russians and Americans who had negotiated the deal were there, including Zhomov. The spies had not escaped him yet. This was his last chance to see the men he regarded as traitors and he would be on the plane with them. The prisoners could feel the tension. Sutyagin would later say he could sense confusion from the men who had until then been his jailors. "They absolutely did not know how to behave with us." Some made their anger clear at the fact that a group of men regarded as traitors were being freed. CIA station chief Hoffman was also on board. He had given each of the four men a bag but told them not to open it yet. The flight was mercifully short. Sutyagin and Skripal sat opposite each other and spoke briefly about family. Vasilenko and Zaporozhsky were more animated as they talked. "Do you think it's a trick?"

Vasilenko asked his former KGB colleague. But by midday, they began their descent.

Their destination was Vienna. Few places could have been better suited for a spy swap. It had been a playground for spies throughout the Cold War. Most famously, it was the site for the iconic film *The Third Man,* written by former MI6 officer Graham Greene. The familiar resonances of the past seemed to echo in everyone's minds. "The Cold War was over, but the scene in Vienna was proof that the old games were alive and well," CIA director Panetta later said. "All that was missing was the sound of the zither playing the theme from the movie *The Third Man.*"

THOUSANDS OF MILES away, the Russian illegals held in New York were also on the move. They were taken by bus from the court late on Thursday, July 8, to a secure area at LaGuardia Airport. Elena Vavilova did not have time to change clothes and remained in her orange prison uniform. They boarded a Vision Airlines red and white Boeing 767-200 charter plane. On the tarmac, some of the field agents who had worked the case said their good-byes to the Russians they had spent years watching.

As they waved them off, some of the FBI team harbored a secret desire. The paperwork had been signed and there was no risk of further incriminating themselves, so the FBI agents hoped the illegals might suddenly open up. They already knew every detail of these people's lives—they had bugged their houses and listened to the most intimate conversations between husband and wife and parents and children, they had read their messages back and forth to Moscow Center, and knew when they were telling their bosses the truth and when they were lying. But what they really wanted to do was ask the human questions as one professional spy to another. What was it really like? Did they have any doubts? "If I have a regret it is that—because I practically lived with the Murphys for so long . . . Richard Murphy in particular—I feel I know him better than some of my relatives, which may say something quite bad about me," says Maria Ricci. But for the Russians, they were heading home to an uncertain future. Being too friendly with the Americans who busted you would not go down well with Moscow Center. The questions were left unanswered.

Ricci handed the Murphys their passports and explained that when they got on the plane they would also get their children's documents, including birth certificates. She reassured them that when they got to Moscow their children would be there. The Murphys thanked her.

At that moment, Anna Chapman piped up. "Can I have my passport back?" she asked the FBI officers. "No, sorry," they said. "Can I have my UK passport back?" she asked. "No, we are going to keep that one, too." "Really?" "Sorry about that. We are going to keep that one." She seemed particularly worried about the impact of the arrest on her British citizenship. Within days of the swap, she was notified via her lawyer that her British citizenship was being revoked and that she could not reenter the country after the swap. She was "particularly upset" at the development, her lawyer said. For all the glamour and trappings of life as a spy she was still a young Russian who wanted to live in the West.

THE FEW DOZEN passengers boarded the Vision Airlines flight. On board the plane were the illegals, some of the FBI team, as well as representatives from the CIA and State Department. They took up only a few rows of seats, leaving many empty. There had been a little bit of competition in Washington about who would get a seat on the flight and witness the end of the show. But those who did make it would later reflect that it would all seem almost anticlimactic. "It was a bit like after a football match where you are banging your heads against the enemy for so long and then it is all over and you shake hands at the end of the game in the middle of the field," says Alan Kohler, who was one of those on board. The illegals—still in prison clothing—were exhausted and somber. "I was exhausted and slept almost the whole flight," Vavilova recalls. "We could comfortably stretch on the seats. Business class meal was a treat after the prison food." The plane took off around 9:30 p.m. for the overnight journey.

Murphy was one of the most personable on board. He was curious as to how the FBI had gone about investigating them. "Did you really have microphones in my house?" he asked. Kohler said they did. Others were quieter. Zottoli withdrew into himself and became introspective, trying to process what had happened to him in the past two weeks. Kohler and another agent tried talking to him and asking him what it was like living in the United States as an illegal. He went quiet. "I always knew this day would come," he said. But, he explained, the longer he lived in the United States, the less he had thought about it. "I could have done this forever. But we lost this one and you guys won. That's just how this one went," he said.

The Russians had sent someone over to escort the team back. Ostensibly this may have been for their welfare, but it would also have been to keep an eye on them and make sure no one had any last-minute changes of heart. The minder brought with him all the recent newspapers and he handed them out. The illegals were amazed to find the papers were full of their stories. After days stuck in detention, they were now beginning to understand that their arrest was major international news. Anna Chapman was not making herself popular, and she was not in a good mood. The newspapers in recent days had focused heavily on her. "Russian Spy Babe's Hot Affair," one New York *Daily News* headline read, " 'Anna Chapman was kinky and great in bed,' says ex-husband

Alex." The article began: "She may have been a true Cold Warrior but she was red-hot in the sheets." There were references to whips, nipple clamps, and their joining of the mile-high club. What really annoyed her was that there were pictures of her naked.

She complained angrily about the stories to the FBI. She began pushing the pages with the pictures of her naked into the faces of an FBI agent, telling them to look at how outrageous it was. I've actually seen them, one of the FBI officers said. Everybody's seen them, he added. "She was the most selfish, self-centered person I have ever heard about or seen or met," one agent recalls.

Once the flight landed in Vienna in the morning, the illegals prepared to depart. Some of the parents thanked the FBI for how they treated them and their children and then they began to get off. The FBI team stayed on board. The swap had been synchronized carefully. The two planes landed within minutes of each other and taxied to a part of the tarmac as far away as possible from everything else at the airport, almost nose to tail with each other.

The illegals never touched the ground. "We went down the ramp and saw people go down the ramp opposite," Elena Vavilova later recalled. A bus then transferred each group to the other plane but they never met their opposite numbers. Sutyagin was given instructions as he disembarked from the Russian plane: do not say a word, do not look around, do not look at the other prisoners, do not make any gestures. "It took 40 seconds. We stepped on the ladder and we were aboard the American plane," Sutyagin later said. Vasilenko walked past Zhomov, his former tormentor. "Nice try, motherfucker," he said. Freedom was starting to feel real. But there was still an air of menace. "Watch out," Zhomov said to Zaporozhsky, according to one person. In that brief exchange lay a warning that perhaps the swap was not quite the end of the story in the eyes of the Russians. The American spies would only later realize that Zhomov had been a few feet away from them. But they never saw him, their nemesis remaining as elusive as ever.

The four Russians boarded the US plane. Doctors gave them a quick examination. They were not in great shape. They smelled like they looked. "He was half the man he was when he had left," one person who saw the once athletic and now emaciated Zaporozhsky would say of his state. The contrast with the illegals going the other way could not have been greater.

THE RUSSIAN PLANE took off for Moscow at 12:38. The American one left seven minutes later.

THE FOUR COMING out of Moscow were still decompressing and coming to terms with their sudden change in fortune. They were hungry and exhausted and trying to take it all in. The bags Hoffman had given them were opened. Inside were clothes and a small bottle of whisky. It was time for a toast. Sutyagin seemed unhappy and reflective, still bruised by having to sign his confession. He kept to himself, making it clear he did not want to talk. The problem for Sutyagin was that he was in strange company for a man who had always maintained he was not a spy.

THE PLANE WAS going to make a stop before returning to the United States. It landed at Brize Norton, an RAF base in Oxfordshire. There was no band or big welcoming party, just one official and two soldiers who struggled to get a staircase to the huge plane because it was far larger than those that normally landed at the base. For Skripal, this was his first time setting foot in the country he had been convicted of spying for. Waiting was a small helicopter to whisk Skripal and Sutyagin away. They were taken to the Fort, MI6's training facility on the south coast. When they arrived,

Skripal and Sutyagin were taken to a room. Laid out were shirts, trousers, even socks of all sizes so they could pick ones that fit. Sutyagin picked light colors after the gloom of the prison uniform. A doctor and psychiatrist checked him over. What did he want to eat? He really wanted *pelmeni*—dumplings—but he was sure they would not be able to get the real thing, so he said he was fine. Sutyagin would say he dreamed of returning to Russia—but he was not sure it would be the Russia he remembered eleven years earlier, when he had been arrested. And in the coming days there were also warnings from multiple people that it might not be safe.

The first thing Sergei Skripal wanted to do was eat fresh fruit for the first time in years. There were no thoughts of going back for him—instead he would disappear into a quiet life in a suburban cul-de-sac in the cathedral city of Salisbury. What could be more peaceful?

THE VISION AIRCRAFT continued on from the United Kingdom over the Atlantic. Vasilenko was the most talkative and cheeriest, describing what he had been through. After a while he was so tired he signaled he wanted to sleep and headed to the back of the plane. Two agents helped lift up the armrests so he could stretch out. They grabbed a pillow for him. It was the best they could do, but for most people sleeping across a row of economy seats is akin to low-level torture. But Vasilenko then said something that made them realize what he had gone through. This was the first time he had slept on something that was not concrete for five years. And then he promptly fell asleep for a few hours. He was awake by the time they flew down the Atlantic Seaboard past Manhattan. There was the Statue of Liberty. Vasilenko stared out the window as if he could not believe he was back.

The plane landed at five thirty in the afternoon at Dulles. A fleet of SUVs was waiting to take Vasilenko and Zaporozhsky to their new lives. After a long and painful hiatus they were going to slip into American suburbia—as anonymous as the spies who they had just been exchanged for in Vienna. For those who had fought the long battle there was a brief sense of triumph. At the CIA an all-hands meeting of the division behind the operation was called. Hundreds of staff were told of the success in getting agents to whom they owed a debt out of Russia. But there was some tension about how far to publicly rub the Russians' noses into the dirt. The Department of Justice, and also the FBI, wanted to make the most of the success of the arrests and swap. But White House officials ordered the DOJ to write the shortest possible press release rather than one trumpeting their victory. That annoyed some.

In public the Obama administration was doing its best to minimize the seriousness of the case in order to avoid damaging the reset. The message was that this was an incompetent group of low-level spies who had never got hold of any classified information. The swap was the easiest way to deal with this and it was time to move on. "We believe that this is a successful resolution to the situation, one that was handled quickly and pragmatically, and we're now looking to move forward in addressing our agenda that we have with Russia and focus on resetting the relationship," the State Department said at their briefing on the day. A White House official on the same day told the *Wall Street Journal* that a major concern for officials over the last few weeks was to "ensure this was not going to have a negative impact on the strategic relationship we have with Russia." This was the message from the administration. A strange Cold War throwback had been initiated because of some old-fashioned Russian spying. This had been dealt with and it was time for everyone to put the whole thing behind them.

But seeing the whole illegals episode and the resultant swap as simply a throwback would prove to be incorrect. It was a mistake that would have consequences.

23

Anger

INSIDE THE PRIME minister's office in Moscow, Vladimir Putin threw papers up in the air. He had just been told of the arrest of the illegals and he was "livid," according to one person with an understanding of the Russian leader's reaction. The illegals program was the pride of Russian foreign intelligence. Now he was having to watch its prized operations against its main adversary dismantled. And it was not just their exposure. It was the humiliation. People were supposed to be scared of Russian spies abroad and proud of them at home. Instead everyone was laughing at them. "Four spies for ten clowns," was how the Russian media described the swap. "A staggering success in the fight against world espionage: Russians exchanged for Russians," another newspaper wrote sarcastically. Alongside public humiliation came professional embarrassment. The FBI had owned these illegals for a decade. Russian intelligence had invested huge effort in placing them undercover, but the Americans had been watching their every move from the start. This cut to Putin's own self-identity as the master spy. "How bad does that make Putin feel," a US spy from the time asks rhetorically. "He definitely was really pissed when it happened." Putin had defined himself as the man who had ended the humiliation of Russia after the terrible decade that preceded his rise to power. The arrest of the illegals cut through to the core of that image he projected. He and his spies had been humiliated.

THE FALLOUT IN Moscow was incendiary. President Medvedev had just been in Washington rolling up his shirtsleeves and eating burgers with his new friend Barack. But the minute he flew out of North America, Russia's nose had been rubbed in the dirt. He had been made to look stupid. US officials did their best to explain that this had not been a deliberate attempt to play him. But, of course, they could not explain that the real reason for the timing of the arrests was to get their agent out of Moscow and that they had actually delayed the arrests to reduce the humiliation. So it left the Russians thinking the timing had been deliberately provocative. For Vladimir Putin, there was a lesson that he would not forget. Medvedev had been played. Russia had been humiliated. His handpicked successor was not up to it. The Americans could not be trusted. The spy game would go on. But it was much more than a game. And the stakes would be raised.

ON THEIR RETURN the illegals were taken to the forest at Yasenevo. They were subject to a "debriefing" as the SVR sought to understand what exactly had happened and if there could have been any further compromises. It was not a full interrogation, but reports suggested lie detectors were involved, as they were kept on a compound for a few days with no mobile phones. For the Kremlin there was a decision. How should they treat the returnees? One former US official says Putin was personally upset with the illegals. They had grown too comfortable living in America on SVR money, even arguing about who owned their houses. He was particularly upset, a former official says, at the idea that the illegals' children were American citizens since they had been born there. The illegals had failed. They had been caught, admittedly not due to their own mistakes, but still, their sometimes sloppy tradecraft and their complaints about the SVR were now on public display thanks to the FBI's indictment. There were no triumphs and no achievements to convince people the investment had been worthwhile. But at the same time, there was another side to the story. These were illegals—so storied in Russian culture. What message would it send to punish them further? To acknowledge failure would be to look weak. And so, in the end, Putin personally signaled that the illegals were to be welcomed back as heroes. He did that by singing a song.

WITHIN DAYS OF their return, the ten found themselves meeting Russia's prime minister. It was a "pleasant" meeting, Elena Vavilova said; "he has a very good understanding of our profession and respect for it." Putin also spoke about the encounter in positive terms. "We talked about life. We sang, not karaoke, but to live music," Putin told reporters soon after. He made a point of mentioning the song they had sung together. "Where Does the Motherland Come From" was the theme song to *Sword and Shield*—a 1968 TV series about an illegal who infiltrates Nazi Germany (the answer to where it comes from is "From the oath that you swear to her in your youthful heart"). The series was one of Putin's favorites and played a part in his desire to join the KGB. Now he was singing it with real illegals. He made a point of praising their virtues and those of all illegals. "Just imagine," he said. "First you have to master a language as if it were your own, think in that language, talk in it, fulfill the task set in the interests of your motherland over many years, suffering daily dangers." Putin was playing to a traditional narrative. Russia could still go head-to-head in espionage with the United States. The die had been cast. There would be heroes—the illegals—and there would be villains—the traitors. These brave Russian heroes had been undone only by treachery. "It always ends badly for traitors: as a rule, their end comes from drink or drugs, lying in a ditch," Putin said.

Putin made these comments during a trip to neighboring Ukraine. Just before talking to reporters he had spent time with a group of Russian bikers calling themselves "the Night Wolves." The group were fierce Russian nationalists who were holding a convention in Crimea, home to a major Russian naval base even though it was in Ukraine. Ever the action man, the Russian prime minister got on a Harley-Davidson three-wheeler at one point to ride alongside them. Dressed all in black, he explained that motorbikes were a symbol of freedom. The iconography of a Harley-Davidson may be quintessentially American but what few appreciated at the time was that on this visit, Putin was already revving up for his return to the presidency on a new agenda. Glorifying spies, vilifying traitors, and renewing nationalism would fuel his return. Vladimir Putin was not done in restoring Russia. Less than four years after his visit, the Night Wolves would be patrolling the streets as Russia occupied Crimea.

Putin added one more intriguing line when he was talking of traitors lying dead in a ditch. "Recently, one of them ended up like this. And it is not clear why." Whom was he referring to? On June 13 Sergei Tretyakov, the former colonel in the SVR who had been working for the CIA out of the Russian mission to the UN at the same time as Poteyev, was eating at his home in Florida. His wife had baked his favorite chicken pie. They were sitting at the table discussing their plans for July Fourth when he got up from the table and then fell. She dialed 911. Paramedics were there within minutes. They tried to do their best. But it was too late. He had died at age fifty-three. The news had only emerged on July 9—the very day of the spy swap. His wife had wanted to keep it quiet to avoid giving Moscow the pleasure of knowing he had passed away. She explained that he had a heart attack after choking on a piece of meat. He was a smoker and drinker, so this was certainly possible. But could there have been more? He had lived openly under his name, only enjoying protection when traveling outside the United States on various trips to brief foreign intelligence services (including in the United Kingdom). He and his family believed he was untouchable in the United States. "Russians would never dare to do anything to Sergei!" his wife later wrote. "If ANYTHING happened to my husband because of Russians, Russia would be excluded from the international community. Sergei was too well known. This would be much worse than Mr. Litvinenko's scandal. My husband died of a heart attack."

So what did Putin mean by his reference? There were some who thought he was hinting that this had been murder. His language, though, was ambiguous. An autopsy confirmed the cause of death. Many US officials working at the time believe there was nothing sinister. However, one senior US official does say that in hindsight it would have been better if more checks had been made at the time to be sure what killed Tretyakov. The same official says that recruits at Russian intelligence training academies are told about Tretyakov's story in a different context. They are told that the FBI will try to recruit you and then when they are done with you, they will kill you. And the story—even though false—is believed by many. The FBI knows that because sometimes when its officers have pitched to a Russian officer serving in the United States—the next generation of Poteyevs or Tretyakovs—the Russians have told this story.

Even if the Kremlin was not behind his death, then there were benefits to having people think it might have been murder since it would act as a warning to others. Tretyakov had been the number one traitor on the list for the Russians, one American official says. But now that he was gone there would be another person to take over the top slot. And that would be the man who betrayed the illegals—Alexander Poteyev.

IN THOSE FIRST few weeks and months there was speculation in Russia as to how the illegals had been betrayed. Tretyakov's defection back in 2000 was thought by some—wrongly—to be the cause. There were reports that the illegals had been interviewed to see if there was a double agent among them. There was even talk that perhaps Metsos had been the CIA mole, since he had mysteriously disappeared. A name of the betrayer appeared in the press but it was wrong (it was actually the name of the person who had provided Hanssen's file). But of course, inside the SVR they must have known the truth from the start. After all, Alexander Poteyev had not turned up for work on Monday morning at Department 4 of Directorate S—just hours after the illegals he handled had been arrested.

This was a full-blown crisis for the SVR. The arrests were described as one of the greatest failures ever. There was talk that perhaps the SVR should be merged with (meaning swallowed up by) the increasingly powerful FSB, with senior figures, including Putin, said to believe that the breaking apart of the KGB into separate groups at the end of the Cold War had been a mistake. The FSB had already been given the power to conduct operations overseas in 2003 and some suspected it of leaking details of how poor the SVR's security had been in order to achieve their ends (which included gaining power to oversee internal security within the SVR, the only domain its reach did not stretch to). The fear among some Russians, though, was that putting all intelligence into the hands of a single organization would bring back the KGB in a new form. The CIA picked up intelligence that the reset might be over and that Fradkov might lose his job. Whatever baby steps there were toward intelligence sharing were stopped in their tracks. "It was all switched off," says one intelligence official serving at the time. The view from the hawks in the CIA's Russia House was that cooperation wasn't worth the effort anyway.

The failure also raised to public debate an existential question about the whole practice of sending illegals overseas. Was it worth it? All that investment and training seemed to have produced nothing but embarrassment. The fact that the US administration had deliberately played down the importance of or threat from the illegals—emphasizing that they had not stolen any secrets—now played back into Russia. One estimate was that the cost of training and supporting illegals and also replacing the communications system that the FBI had penetrated and compromised was $50 million. "What damn illegals? They haven't been needed by anyone for 50 years. Someone just had to report to their boss that we have intelligence operations," an anonymous former SVR officer told the *Novaya Gazeta* newspaper. The whole enterprise was an anachronism, a hangover from the Cold War whose only purpose was for spy chiefs to be able to impress their KGB-trained leader that they had one over on the "Main Adversary" by having agents secretly in place in their country, argued Andrei Soldatov, the leading independent commentator on Russia's spies. "Still stuck in the past, Putin views this superpower rivalry much in the same way he wants Russian athletes to get more medals than the Americans at the Olympic Games," he wrote. Others were also skeptical. "Perhaps, Russia will save some money now after it stops spending money on illegal intelligence," mused Alexander Lebedev, a former KGB officer who had moved to Britain and become the owner of the London *Evening Standard* newspaper. This questioning was one of the strategic effects that the Americans hoped to accomplish by public arrests—the humiliation of defeat and the inevitable questions in Russia over expense would, it was hoped, make it harder to sustain the illegals program going forward.

Old SVR hands did their best to defuse this line of attack. "Even if we knew the final amount it would be trifling in comparison with the huge profit which we could get from the work of intelligence agents in the future," explained Yuri Kobaladze, a former KGB and SVR officer. "Imagine that 11 people, embedded, settled in, and absolutely naturalized, go to work at the CIA or the Pentagon in a couple of years; Anna Chapman marries a Senator; and Russia starts to receive top secret information. This means millions!" Here perhaps was the most succinct explanation of what SVR officers had in mind for the illegals to achieve. But the time of the old illegals was passing. There would be new ones to take their place.

FOR THE ILLEGALS themselves, there was now a new life to come to terms with and some complicated relationships. The first time Donald Heathfield—now Andrey Bezrukov—met his sons after release, he was still in his orange prison uniform. "It was already late, but we talked deep into the night," Alex Foley later said. "Slowly, we began to discuss their past. We discovered a confusing patchwork: stories from their childhood having been real, but taking place in the Soviet Union, not in Canada or Europe. In general, my parents were open and spoke about their hidden lives and identities. With regards to their methods or missions, my brother and I were smart enough to not ask questions where we knew we would be better off not knowing the answers." Bezrukov would say he spent the first month explaining his life to his sons. "In the end, they had an understanding of why we made a certain choice in life," he would say. "They understood us," his wife, Elena Vavilova, said. "They understood our choice and accepted it." What had it been like to be born in the Soviet Union, work in America, and then return to Russia? "The fact that I left the country, which was called the Soviet Union, and returned to one called Russia, did not affect me in any way," Bezrukov would say. "For me, this is one country. My country."

For the children, though, Russia had not been their country and it was a deeply disorienting time. Parents had to explain to their sons and daughters that they were not who they thought they were. Their lives had been built on deceptions that had now been exposed. In one case, according to multiple sources, the children of one illegal family, likely the Murphys, were left alone in a room in a building used by Russian intelligence. Their parents then walked in. But now they were in full Russian military uniform. That was thought to be the best way to convey the truth of who they really were. How the children reacted is not known, but it would be hard to imagine anything other than shock and bewilderment. One thing that convinced the Foley boys of the truth was when their mother showed she could take down Morse code at twenty groups a minute with almost no mistakes.

The illegals and their families were taken on a tour after their return. They went to St. Petersburg, the Black Sea, and Siberia. It was a chance to reacquaint themselves with the country they served or, in some cases, to see it for the first time. They would have made a motley crew, retired Juan Lazaro, Anna Chapman, and some bewildered kids.

But what were the illegals to do now? Could they go back to work for Directorate S? That was not possible. Trust was always the issue with those who had lived abroad. When the illegal Rudolf Abel returned after his swap, he complained of being a "museum exhibit" and not being used very much. That was because the KGB feared he might have been turned in American detention. His bedroom was surrounded with listening equipment in case he talked in his sleep and revealed the truth. "He was sore as hell," one friend said of his treatment. The same risk meant that the former illegals could no longer work on active operations although they were all decorated with awards and medals.

Eventually, jobs were found for all the illegals—and good ones at that. They had to be successes, after all. Michael Zottoli, now Mikhail Kutsik, went to work for Gazprom and in 2018 met EU officials to talk about regulatory issues for the industry.

Donald Heathfield had the most prominent profile. He maintained a LinkedIn page under the name "Donald Heathfield aka Andrey Bezrukov," with a picture of him smoking a big, fat cigar. (His wife, Elena Vavilova, maintained a LinkedIn page with a rather more playful picture listing her as having become an adviser at Norilsk Nickel, a Russian mining company, in early 2011.) Heathfield got a job as an adviser to Igor Sechin, the president of the energy giant Rosneft and one of Putin's closest allies. He also lectured at the Moscow State Institute of International Relations with a distinctive accent that veers between Moscow and Boston. "Class with Andrei Bezrukov, a former spy within the United States, was one of the highlights," one visiting student from Seton Hall University in New Jersey said. "The class was very informative."

What's it like to be a spy? Bezrukov was asked by a class of Russian students in May 2018. "Just watch *The Americans*—I'm not kidding," he told them, although a show of hands revealed most had not seen the award-winning TV series about illegals set in the 1980s, and which had been inspired by his own career. Bezrukov had watched it in Moscow and was a fan. "The result is quite close to reality," he said, "though without the killings and the wigs. The creators of the series were able to show the atmosphere, and the inner feelings of illegals, and the difficulties, including personal ones, that one has to face, and the personality that is required to do such work. Many moments are conveyed accurately, honestly, and professionally. My wife and I did not expect that the American creators of the series would be willing and able to show the characters of the spies so deeply and unbiasedly, even with sympathy."

"The producers have captured well the atmosphere of the eighties," Vavilova told me recently. "The illegals have a human side with believable emotions and problems of their daily life. However, violence and disguise are never part of the professional work. This way the spies would have not been able to last for a long time. I believe those scenes were necessary to keep the attention of the viewers. . . . I can hardly believe that real illegals could have met their handler inside the US and they would have immediately moved to another house after discovering that an FBI agent lives just across. But again, real life and movies are not the same thing."

Ironically, many US officials who worked on the case struggled to watch the show. It was too close to home and they were wary of the attempts to portray the Russian spies too sympathetically. Told about Heathfield's view of his own fictional portrayal, one person who watched him while he was undercover comments wryly: "I guarantee his sex life was not as interesting as theirs."

Watching the show, Bezrukov's brain would suddenly switch back to English language. And the family still spoke together mainly in English five years after the swap, he said in an interview. It took nearly five years for the pain and for a kind of nostalgia for the past—to hear the language and live in America—to disappear. "What I regret most of all, is that I did not have an opportunity to complete my mission," he said. He and Elena had to rebuild relationships. "We had lost all our friends. I have lost my father. But some people who worked abroad much longer than we did, came back to emptiness. No parents, no friends. Nothing," he later said. The greatest toll had been on his family rather than himself. "Nobody asked my kids whether they wanted to live the life like that. Or what are the consequences of being the sons of spies or something like that or parents who didn't see us for long, long years. In fact, we create more problems for the generation before or generation after. That's the biggest human drawback in this. Otherwise, it's the job I loved." His wife also felt she had done the right thing. "There are certainly no regrets that I made this choice back then," she said in Tomsk years later. "Were I offered to start my life again, beginning with my student years, I think I would have made the same choice. . . . For me my profession has become my life." She kept the orange prison uniform she was swapped in.

NOW IT WAS their children's turn to be strangers in a strange land. They were Russians by blood but not by culture or upbringing. All of Alex's and Timothy's plans for their future were torn in shreds and they found themselves living in Moscow, far away from friends and their former life, unable to speak the language. There could not be anything other than some degree of resentment at what they had been put through by no choice of their own, even though the family bonds would ultimately prove strong. The boys' nationality was contested as they entered a bizarre netherworld. It turned out that unpicking a persona for the sons of illegals was nearly as complicated a process as building one for their parents. The United States said Alex's passport and citizenship were no longer valid. When they heard talk the boys wanted to come to America, a senior FBI official and his deputy met with two counterparts from the Russian embassy at an Italian restaurant near FBI headquarters. No one ate anything. The Russians folded their arms and listened as the FBI officials explained that the United States did not want the children in the country and the Russians should tell them to stop trying.

Alex found even his name was no longer his own. He was given a Russian passport under his new name: Alexander Vavilov—taking after his mother. He applied for a Canadian passport using his old birth certificate. He was told he would have to amend the birth certificate to reflect the change in his and his parents' names and provide DNA evidence to establish his relationship with them. Then, if he wanted a passport in the name Foley, he needed to legally change his name back after the Vavilov birth certificate was issued. With a new birth certificate, he tried to get a Canadian passport and was told he needed a citizenship certificate. A student visa to go to Canada was canceled three days before he was due to leave. His application to study in France—made on his Russian passport—was rejected. Visa applications were also rejected by the United Kingdom, Spain, the Czech Republic, Australia, and Hong Kong. He launched legal action to try to get Canadian citizenship, the country to which he felt he belonged. "I am essentially exiled," he has said.

His brother Timothy felt out of place in Russia and wanted to leave the country. In September 2011 he went to London to study at the Cass Business School. He applied to renew his Canadian passport and was informed he needed to receive a citizenship certificate. He was then told that he was not entitled to one because his parents had been working in Canada as employees of a foreign government at the time of his birth. He was also told he needed a new name and became Timofey Vavilov. "I feel I have lost touch with my previous self," he said. He found Russia difficult. "I have felt out of place and a foreigner," he testified. Like his brother, he said he felt Canadian and fought in court for the right to live there.

Timothy continued to dispute the claim he had been groomed to take over from his parents. "These allegations are not true. It has been stated by the FBI that for over 10 years my home was bugged, however no evidence of my involvement has ever been presented. . . . I do not have any knowledge of 'sensitive information' related to my parents or their work as they have not divulged any information to me," he stated in an affidavit for the court cases. The boys' relationship with their grandparents remained confused. "We don't have a very close relationship," said Elena's mother nearly a decade after the swap. "They are polite, of course, and accept us. The elder one is a little more gentle. But the younger one . . . His mother tells him to call, talk, congratulate us sometimes, and he replies that for him the grandparents are strangers."

THE FATE OF the Murphys' two daughters, younger than the Foley boys, preyed on the mind of some of the FBI agents. "I think about their kids a lot," says Maria Ricci. "We watched them literally grow up. I hope they are okay. I really do." It would have been far more bewildering at their age to be ripped out of their surroundings, but perhaps the lack of adult understanding may have made it easier and led to less questioning of their parents' choices. A friend of eleven-year-old Katie received an email from her in Russia, a brief attempt to keep in touch and reclaim some normality. The smart, well-behaved kids who sang "God Bless America" were technically American citizens since they were born in the United States. Could they come back to America one day as adults? And how would the United States and the FBI view them? As innocents who had been lied to by their parents? Or potentially the next generation of illegals? A decade later, the girls have been able to move on and build new lives as young women, lives that reflect their international upbringing, one of them studying in Europe and able to speak five languages.

VICKY PELAEZ HAD been deported to a country she had never before set foot in, with a husband whose real name she says she never knew. She could not speak the language and found it hard to cope with the separation from her children. She wanted to return to Latin America. Her husband, too, it was reported, made clear he wanted to go to Peru to live as Juan Lazaro. "He doesn't want to stay in Russia," his American lawyer, Genesis Peduto, said just a few weeks after the swap. "He says he's Juan Lazaro and he's not from Russia and doesn't speak Russian. He wants to be where his wife is going, to her native country, where it will be easier for Juan Jr. to visit," the lawyer said, adding that Lazaro put his family first. This again had always been the concern for Russian illegals who did not marry another illegal—that the bonds of marriage and children would supplant those of patriotic duty.

Pelaez says for the first few months she struggled to adapt to her realization that her husband was not who he had said he was. Half a year after the swap, they were still living together and discussing it all the time, Pelaez admitting she had used some "very strong words" that hurt him at times. She maintained she was "collateral damage" in the spy game between the United States and Russia and had never known of her husband's activities. Pelaez went back to Peru in 2011 for her father's funeral and later that year, after an investigation into the use of false documents was dropped, Lazaro received a new Peruvian passport and they both moved there.

IN A BRIGHT red jacket, Anna Chapman smiled and waved as she saw off a group of cosmonauts heading to the International Space Station. Chapman had the most colorful life following the swap. She had already become a tabloid sensation in Russia as well as the United Kingdom and United States before she had arrived back home. But she soon became something more. Her first public appearance in October 2010 at a Russian space launch caused a stir. Her next outing was more revealing. A few days later she posed nearly naked and wielding a gun in a photo shoot for Russia's *Maxim* men's magazine. But there was also a more serious side. She was given a senior role in the Young Guard—the youth wing of Putin's One Russia party. There was talk for a while that she might go into politics.

She made her first TV appearance in December on a talk show in which her patriotism was stressed, even though she spoke fondly of her time in the United States and Britain. The next month she was fronting her own show—*Mysteries of the World with Anna Chapman*. "I will reveal all secrets," she promised in a red and black evening dress, but the only secrets of the first program were about a baby in the North Caucasus with skin marks said to look like Koranic verses. She built her brand with the same gusto she had displayed years earlier, launching her own website, which focused on her charity work rather than her former life in the West.

Chapman had become a perfect symbol, as Russia commentator Edward Lucas puts it, of the "sleazy glitz" of modern Russia, in which sex, spying, business, and politics collide. She had taken the decades-old Soviet cult of the spy and illegal but then added on the trimmings of post-Soviet Russia, with its love of outward displays of wealth and glamour and a splash of nationalism to make up for the passing of ideology. She made a perfect feminine counterpart to a KGB officer turned president who liked to appear with his shirt off to emphasize his masculinity. They were supposed to represent vitality, vigor, and national pride. The cult of the spy had been reborn for a new TV age. It is said that each country gets the spies it deserves. If so, modern Russia would perhaps have a fusion of Anna Chapman and Vladimir Putin.

BACK IN BRITAIN, her ex-husband, Alex Chapman, would be interviewed by MI5 to see what they could learn about his ex-wife's past. There was a brief flurry of interest in the racier details of his marriage as the young man did his best to make sense of what had happened to the girl in the white dress he had meet at a rave eight years earlier. Anna's friends in Russia hit back when he started talking about how she had changed. "She quickly became disappointed in Alex and understood that he was not a man but a rag," one said when explaining why they had broken up. But while his former wife would go on to become a tabloid sensation and TV star, Alex Chapman did not fare so well. In 2016, aged just thirty-six, he died of a drug overdose.

ANNA CHAPMAN ALSO had one more role to play nearly a year after she returned. That was in the trial of the man who betrayed her.

Alexander Poteyev may have been on trial, but he was not physically there—just a ghost in the courtroom. The trial took place largely in secret at a Moscow District Military Court, behind large locked doors with a guard at the entrance. A panel of three judges heard the evidence collected by the FSB. Poteyev was accused—in absentia—of high treason and desertion. Only the final verdict was in public. A few veterans of the SVR turned up for the moment when Poteyev was sentenced to twenty-five years in prison and stripped of the rank of colonel and all his state awards. The star witness was Anna Chapman, whose written testimony was read in court. She said she was sure Poteyev had turned her and the others in. She said only a small group would have known about their activities. She identified Poteyev by photo. She also said that the special password that the undercover FBI agent "Roman" had used was one that was in her personnel file at Moscow Center that Poteyev had access to.

The realization they had been betrayed by a colleague cut deep. "We didn't much like Poteyev himself as a person," Elena Vavilova said later. "For this person, money became more important than serving his country." What would the illegals say to the man who betrayed them? "He had a choice. He made his choice. He should judge himself," she said. Her husband felt the same.

"I would not say anything," Bezrukov said when asked. "In my opinion, he will be rotten for the rest of his life. Betrayal, like an ulcer, if it is in you, it will eat you up. . . . His father was a Hero of the Soviet Union. He betrayed not only himself—he destroyed the memory of his parents." Bezrukov said that his personal impression of his fellow spy had been of a "weak" professional. He may have been welcomed by the CIA and FBI at first, but in the long run they would suck him dry and no longer need him. "Like a squeezed lemon," said Bezrukov. He could barely contain his anger at a man who had negated more than two decades of sacrifice working undercover.

"It was no coincidence that Dante placed traitors in the ninth and last circle of hell," SVR head Fradkov said a few months after Poteyev had fled. The message was more direct from others. "We know who and where he is. Either he betrayed them for money, or else they simply caught him at something or other. But you need have no doubt—they have already sent Mercader after him," one "high-ranking employee" of the Kremlin said—Ramon Mercader was the agent who murdered Leon Trotsky with an icepick in Mexico after he had fled Stalin's Russia. "This man's fate is not enviable. All his life he will drag it behind him, and every day he will fear retribution."

IN DECEMBER 2010, Putin took part in one of his regular TV appearances where members of the public could ask questions. Someone emailed in about the swap: "When speaking about the recent spy scandals, you said that traitors don't live long. As we know from memoirs, the leaders of many countries signed orders to eliminate traitors abroad. The French and the Israelis did it. As head of state, did you ever have to take such decisions?" Putin's reply was long. During Stalin's time, there had been "special groups" to eliminate traitors. But these were now out of service.

Russia no longer maintained such a practice, he explained. "As for traitors, they will drop dead without any assistance because . . ." He paused for a moment. "Well, take the recent spy scandal, in which a group of our undercover agents was betrayed. They were officers, you understand? And the traitor exposed his friends—his comrades in arms whose lives were dedicated to serving their homeland. Just imagine what it means to speak a foreign language as a native tongue, to give up one's relatives and not even be able to attend their funerals. Think about it! A person spends his life serving the homeland, and then some bastard betrays him. How can he live after that? How can he look into the eyes of his children, the swine? No matter what gains a traitor receives for his malice, thirty silver pieces or what have you, he will never derive any pleasure from them. Spending the rest of your life in hiding, unable to talk with your near and dear ones—the person who chooses such a fate for himself will regret it a thousand times over."

The claim that Russia did not kill was a lie. Only four years earlier, Alexander Litvinenko had been murdered in London. But of course, Putin had to maintain that line. What was more telling was the length of his answer and his intensity. Even Russian journalists used to Putin's coarse language noted that talk about swine was unusually brutal. It reflected a deep personal anger. Vladimir Putin was not done with traitors. Nor with America.

24

Still Among Us

It was early evening, the day after the FBI swooped down on the illegals in suburban America, and a New Zealander called Henry Frith was walking down the street in Madrid. A stranger came up to him. This was no chance encounter.

"Have you got just a few minutes to chat with me?" the stranger asked in a British accent.

"I'm sorry," Frith replied.

"I think it's very important that you do," the Briton continued. "I have here in my hand your life. . . . If we do not talk now then I'm afraid there's going to be a big problem for you here in Spain. . . . I work for Western special services, and you work for the Russian special services. I know this is a shock and I'm sorry that I have to do this on the street, but it was the only way I could get to talk to you securely."

The Briton worked for MI6. This was a staged confrontation planned back in London at MI6 headquarters at Vauxhall Cross. As he kept talking, Frith, the supposed New Zealander, kept saying "You're wrong" and "no." The Briton made one final pitch. Become a double agent against the Russians, he offered. "I have the opportunity to make your life a lot, lot better" was the promise. But there was also a threat. "The Spanish authorities are going to come, the police are going to come, there's going to be a big scandal, and your whole life is going to be in ruins." It was a stark choice. Leave behind the life he was enjoying and go back home with his cover blown and career over. Or maintain his comfortable life in the West, always looking over his shoulder.

Henry Frith took a pass on the offer and walked away. Early the next morning he rushed to the airport. He was never seen again.

Frith was another illegal. The real Henry Frith had been born in New Zealand in November 1936 and died a year or two later. The boy had been tombstoned by Directorate S to create a legend for an illegal called Sergei Cherapanov, supposedly born in Ecuador in 1957 (with a New Zealand father and Ecuadorian mother). Cherapanov is thought to have used an encrypted laptop to write messages before copying them onto a secure USB, which he then left in dead-letter boxes that Russian diplomats would clear. One of those diplomats was pictured pretending to urinate by the side of the road while actually taking something from under a rock. He was most likely running agents in Eastern Europe or the Balkans from his Spanish base.

Cherapanov was identified thanks to Poteyev—hence the need for the fast pitch the next day. It is thought that Cherapanov had been under watch for some time, perhaps as much as a year—just one of the leads provided by Poteyev that the CIA had tried to mask as it passed details to allies. When Poteyev fled, the SVR would have realized that Poteyev could have blown some of these names. Fearing they could be withdrawn, the British decided to move fast. Spanish intelligence was aware of the case as well, but it was agreed that it would be better for MI6 to make the pitch. This was sometimes the case with Russians in Spain (like Sergei Skripal years earlier). There was no attempt to reel Frith in carefully or test his motivations. It was a fast and dirty pitch, but it was also a no-lose situation. "You win the gold medal if he works for you," one former British spy explains. "But you still get the bronze if he buggers off."

THE CURTAIN HAD been lifted on Russian illegals thanks to Poteyev, one British counterintelligence official at the time recalls, and suddenly you had a view of what had been going on. This, another spy explains, offered a whole series of "direct opportunities" such as the one in Madrid. In the United Kingdom itself, they could see the travel of illegals through the country although none was based there. In other cases, there would be the chance to make a move.

The operation that busted the American illegals rippled out widely. "It obviously rang a lot of bells around the world," says Leon Panetta, CIA director at the time. There was disruption to the SVR's global network of illegals. One US source claims that as many as sixty illegals around the world were identified. A Russian report after Poteyev's trial claimed he had blown eighty undercover agents and an entire communications infrastructure that had cost $600 million, making it the largest failure in the SVR's history. The German security service said that in the decade after 2006, the EU and NATO uncovered twenty illegals working for Russian intelligence (although it is not clear if this includes the ten in Ghost Stories). But relatively few cases have come to light. In some instances the United States told a country they had an illegal operating within their borders, but, the country was too worried about causing a row with the Russians to do anything. They—like others—may have decided to watch an illegal rather than pounce to see what they could learn. In other cases, of course, an illegal may have been approached, but, unlike Henry Frith, decided to take the offer and turn into a double agent. That is another reason why the details are kept secret. In many cases Poteyev may have provided leads—nuggets of information about someone he overheard—that were useful to work on but not necessarily conclusive evidence. A decade after the Ghost Stories swoop, one Western counterintelligence source says that threads are still being pulled. But the months immediately after the arrests would reveal two facts: first, that the illegals program was under pressure; second, that it was still going but evolving.

BLACK-CLAD OFFICERS FROM Germany's elite GSG 9 police special forces team crept into the small house in Marburg in October 2011. It was early morning but their target was already awake. On the top floor the officers burst into a room to find Heidrun Anschlag busily working a shortwave radio transmitter. On the other end of the line was Moscow Center. Heidrun was so startled that she literally fell off her chair.

Heidrun, forty-six, and her husband Andreas, a fifty-two-year-old engineer, were long-term illegals with globetrotting backstories (and the confused accents to match). In 1984 a lawyer registered Andreas Anschlag as an Austrian born in Argentina in 1959. The fake birth certificate (along with a Mexican driver's license) all came from Directorate S's forgers. A local official is believed to have been bribed a few hundred dollars to make sure everything went through. Heidrun had apparently been born in Peru a few years later. They had been selected as a couple and may have tied the knot a second time to get her an Austrian passport. In the same way that Canada was the jumping-off point for the United States, Austria was for their target of Germany, where they arrived in 1988 in the dying days of the Cold War.

The case was a perfect example of the damage that could be done by illegals if they were not under control in the way those in the United States had been by the FBI. Andreas worked in the automotive industry and traveled abroad frequently—including to the United States, Spain, Portugal, the Czech Republic, and Brazil. He joined policy groups and think tanks and attended conferences on transatlantic relations security. There, like Heathfield and Murphy, he would suggest targets for the SVR from the people he was meeting—including politicians, government officials, and people linked to the intelligence services who were then targeted by the SVR. The couple received about 100,000 euros a year from the SVR.

As well as scouting for potential targets, the couple also ran agents themselves. A Dutch Foreign Office official, Raymond Poeteray, provided them with sensitive documents about NATO and the EU. He had come back from a posting in Hong Kong (where on their departure he and his wife dumped a Korean girl they had adopted) and found himself in debt and living above his means. Andreas Anschlag drove to Holland once a month on Saturdays to exchange documents on a USB stick for money—close to $100,000. "He also passed on sensitive information about colleagues to the SVR, such as their sexual orientation and possible health issues," a court was later told of Poeteray. All perfect for the SVR to work out who else they might be able to subvert.

The device Heidrun Anschlag had been caught using was a high-tech piece of equipment with an antenna hidden in the lining of a laptop bag. The bag held a high-frequency transmitter that sent messages to one of a half dozen or so Russian satellites circling above. On Tuesdays at 6 p.m. Heidrun would plug a decoder into a shortwave receiver. A red light would indicate a satellite was coming into position and a blue light indicated a message was being transmitted securely. The couple had attended a training course in Russia on the decoding program called "Sepal" and an encoding program called "Parabola." They would inform Moscow that material was waiting in dead drops. Line N officers would head out and empty them.

YouTube's platform provided another novel way of communicating. The couple and the SVR created accounts a month apart in early 2011, which commented on videos, mainly about the footballer Cristiano Ronaldo. "It's a very nice video and the song is also very good," they wrote. This was answered by the SVR account, called Cristianofootballer, saying, "He runs and plays like the devil." This, German investigators believe, was a means of communicating by hiding in plain sight amid all the noise on the world's largest video platform. The comments included a sequence of punctuation marks that could be turned into numbers, which would then refer back to a pre-agreed message. This was the next step on from the famous number stations—public radio broadcasts to spies and illegals in code that have been used by the Russians (as well as other countries) for decades and can still be heard.

A tip-off that came from both the Austrians and the Americans led the Germans to the Anschlags. The Russians seem to have been aware an investigation was closing in—perhaps because they were sensitive after Poteyev's arrest—leading to discussions about returning to Moscow. The SVR was concerned about communications equipment falling into Western hands. Knowing that time was running out, the Germans moved in October 2011. The Anschlags served relatively short prison sentences after an initial attempt at another swap—for a CIA agent in Russia—fell apart. Their daughter became another person who suddenly realized she had Russian parents.

ONE CONNECTION IN these cases is that Poteyev may have been able to tell the United States about the use of Latin American cover identities. That would also link to a Belgian investigation launched in 2010. A man claimed to be born in Uruguay of a Belgian father and British mother, the identity stolen from a British child who died young. His wife was supposedly born in Ecuador to a Belgian father and Ecuadorian mother. Their identities had been created between 1990 and 1992, when their names were inserted into the Belgian population register. The man later requested a passport at the Belgian consulate in Rome, the woman in Casablanca. A Belgian consul in Morocco was accused of having been recruited, including through a relationship with an SVR agent, to approve false identity documents (part of a network of consuls compromised by Moscow). The couple moved to Belgium and had children. Belgium, home to the EU and NATO, has long been a prime target for Russia's spies, and the pair were busy making contacts and passing back information before they went to Italy and then disappeared.

MI5 SUFFERED ONE bruising experience in the United Kingdom immediately after the Ghost Stories arrests, involving a young Russian woman who was inevitably compared to Anna Chapman. On August 8, 2010, a month or so after Chapman hit the news, Ekaterina Zatuliveter was stopped at Gatwick airport. MI5 officers took her to a nearby hotel for the first in a series of interviews. A few weeks later a sensational headline hit the British press: "MI5 quizzes MP's aide on spy links." Zatuliveter worked for a member of Parliament who had previously sat on the Defence Select Committee. "MI5 made its move weeks after a Russian spy ring was uncovered trying to infiltrate policy makers and financial circles in the US. Security sources confirmed that the intelligence agency is investigating whether 'sleeper cells' are also active in Britain," it was reported at the time. In December the government moved to deport Zatuliveter on the basis she was a threat to national security.

Zatuliveter was born in the North Caucasus and studied at St. Petersburg University. MI5's assessment claimed that the FSB and SVR had a substantial presence at the university, including at the School of International Relations, where she studied in order to monitor foreign students and recruit Russian nationals. During her time in Russia she had a brief relationship with a Dutch diplomat based in Moscow. She also met Mike Hancock, an MP with an interest in Russia, during a parliamentary visit in 2006 when she acted as a chaperone for delegates. He invited her back to his hotel room but she refused. But a sexual relationship began the following month. She then came to the United Kingdom to study and eventually began working for Hancock in Parliament. Next, in 2010, she met a fifty-year-old official who held a significant position with NATO in Moscow. Zatuliveter engaged in a series of flirtatious texts with him that eventually led to an affair. "The Security Service assesses that Zatuliveter's sexual involvement with two significantly older men with access to sensitive material and influential positions is consistent with the activities of an agent working for the RIS (Russian Intelligence Services)," MI5 formally concluded. The view of MI5 was that she was an agent rather than an illegal or officer, most likely recruited before she came to the United Kingdom. "I am not suggesting that in this case there was a honey trap in terms of their being an entrapment in order to embarrass or blackmail. The point that is being made here is that the Russian intelligence services still use sexual relationships as part of their operations," an MI5 officer said in evidence. "Our case is that the beginning of their relationship may well have been directed by the Russian intelligence services," they said of her and Hancock, while conceding there could have been genuine feelings as well.

But Zatuliveter decided to fight deportation. In a remarkable court battle—held partly in the open and partly in secret—her lawyers challenged the MI5 case. The press drew the analogy with Anna Chapman. "There are also similarities in that they both happen to be young women, that is a similarity, and that they are very adept networkers," an MI5 officer testified. The case was different, though—there was no proven contact with the Russian intelligence service. Her diaries suggested she was genuinely interested in the men she was involved with. MI5 officials doubt if Zatuliveter was sent out as a spy—but believed that once she was spotted as having interesting contacts, she could have been approached by the Russian intelligence services. But the court found she had not been tasked to seduce Hancock or the NATO official. Their conclusion was that "at least on the balance of probabilities" she was not a Russian agent. Zatuliveter won the case. It was a blow to MI5's spy hunters.

It was not the only contact Hancock had with a Russian woman. In the late 1990s he met another young Russian—Ekaterina Paderina. Paderina had arrived on a student visa in the late 1990s and in 1998 married a retired merchant seaman more than twice her age. The marriage lasted a matter of months. She was served with a deportation notice but went to Hancock for help to remain in the country. She succeeded. Paderina was allowed to remain and in November 2001 went on to marry a millionaire insurance company director. He was not particularly well known at the time. His name was Arron Banks. He was increasingly politically active in the coming years and eventually would become the most public bankroller of the campaign for Britain to leave the European Union. Paderina drove around in her husband's Range Rover with the number plate "X MI5 SPY." She and her husband have always dismissed any accusations she is a spy.

But what of America? There was only the briefest of stand-downs for Russian intelligence in the wake of the arrests. And you might assume the SVR's sending of spies under illegal cover would be at the very least paused. That was not the case. It only took weeks for the Russians to be sending more spies over—albeit under different cover.

. . .

THE ILLEGALS HAD barely unpacked their bags in Moscow when Evgeny Buryakov arrived in New York in August 2010. He was a banker in his mid-thirties who was about to become the deputy representative of Vnesheconombank (also known as VEB), the Russian state development bank. With a wife and two kids, he lived a quiet life in a modest house in Riverdale in the Bronx. He was code-named "Zhenya."

The FBI's New York field office would open an investigation—with Maria Ricci supervising—that eventually led to the indictment of three men. It was the arrival of one of them, Igor Sporyshev, in New York on November 22, 2010, that started things off. In his late thirties, he had been posted as a Russian trade representative, but his father had been a senior figure in the KGB and then the FSB. That gave him the profile of a potential spy and the FBI began to investigate. Another of the three, Victor Podobnyy, was in his mid-twenties. He arrived on December 13 as an attaché at the permanent mission to the UN. Both, the FBI would establish, were undercover SVR officers.

The banker Buryakov was the most interesting figure. He was an undercover SVR officer working without diplomatic immunity. But he was not hiding his Russian nationality. Rather he was working under what Americans would call "non-official cover" as a banker. In that sense he was closer to the Anna Chapman model but was a fully trained officer rather than a True Name or Special Agent illegal trained up on the job. He worked not for Directorate S but Directorate ER. This focused on economic intelligence. Its priorities were information on sanctions against Russia as well as attempts by the United States to develop alternative energy resources, such as fracking, which undermined Russia's economic leverage. Because this team worked for a different directorate from Poteyev, the SVR would have hoped they were not compromised by his departure.

Buryakov had already spent five years in South Africa from 2004 as a banker and undercover SVR officer (even his boss had not known his true role until an SVR officer blurted it out over dinner one night). Buryakov's cover meant he could not visit the SVR's *rezidentura*. Sporyshev and Podobnyy were carrying out the standard intelligence tasks of trying to recruit New Yorkers as sources but were also acting as handlers for Buryakov, transmitting his intelligence back to Moscow Center. Buryakov was careful not to discuss anything intelligence related with the others over phone or email, instead meeting face-to-face.

The FBI began surveillance of Buryakov in March 2012 and watched nearly fifty meetings with Sporyshev. The meetings would come after a quick phone call in which it was explained that some ordinary item like a ticket, book, umbrella, or hat had to be handed over. A dozen times they talked about handing over tickets, but they were never seen attending an event that required one.

There were other slipups. On May 21, 2013, Sporyshev called Buryakov on a tapped phone line. He needed help urgently. A Russian news agency had the chance to talk to someone at the New York Stock Exchange and Sporyshev needed the banker's help with questions to extract the best intelligence. Buryakov said they should ask about exchange-traded funds, including mechanisms of their use for the "destabilization of markets." Were the Russians simply worried about the possibility? Or were they trying to research how to do it? At least one senior US intelligence official believes the Russians were looking at ways of creating their own "flash crash" of the markets if they ever wanted to.

Buryakov was providing intelligence on a major potential multibillion-dollar deal in which the Canadian aircraft manufacturer Bombardier would make planes in Russia. Buryakov traveled to Canada twice to attend conferences using his banking job as cover and learned that Canadian unions were unhappy with the plan. So he came up with a proposal to influence the deal. It would involve calling the resources of the SVR's Active Measures Directorate to find some way of pressuring the unions. This showed Buryakov's value—his cover as a banker allowed him to move in business circles and gain useful intelligence, which in turn could be used to actually influence international business deals to Russia's advantage.

In April 2013, Sporyshev told Podobnyy he was trying to recruit two women. Preferred targets seemed to be young women who had just graduated. He explained to one woman, who worked for a financial consulting firm, that he wanted her help to find material that was not available in open sources—perhaps from specialist journals but also from people who discussed the subjects behind closed doors. But he was having more trouble with another woman. He seemed to want to be some kind of male honey trap but it was not quite working out: "In order to be close you either need to fuck them or use other levers to influence them to execute my requests." The FBI approached the two women, who explained that Podobnyy had tried to ingratiate himself and gain information from them.

The FBI had the SVR's New York *rezidentura* wired thanks to a clever operation. An undercover placed in Sporyshev's path posed as an analyst from a New York–based energy company. She provided the Russian with a set of binders that were supposed to contain industry analyses. She explained they would be missed eventually so they needed to be returned, but she could then get more. Inside each set of binders were covert recording devices. Sporyshev took these binders into his office in the *rezidentura*. When they were returned to the undercover, the audio could be transcribed and translated. And by then a new binder was in place. There would be hundreds of hours for Ricci's team to work through.

There was a particularly amusing conversation between Sporyshev and Podobnyy on April 10, 2013, when Podobnyy complained that life as an undercover spy was not what he imagined when he first joined the SVR. It was not even close to James Bond, he moaned. Even Russian spies seemed to reference their job against British intelligence's most famous fictional spy and complained when reality did not match up. He knew he would not be flying helicopters but he had hoped at least he would be undercover. Sporyshev agreed. "I also thought that at least I would go abroad with a different passport," he said glumly.

A concern that they were not "real" spies was a recurring theme. The pair, a few weeks later, on April 25, 2013, discussed their envy of illegals. "Directorate S is the only intelligence," Podobnyy said.

"It was," Sporyshev said. The reason for him using the past tense, it seemed, was that Directorate S appeared to have fallen on hard times since the 2010 arrests.

"I don't know about now," Podobnyy said, half agreeing before correcting himself. "Some things remain, like the Middle East, Asia, not everything has fallen."

Parts of Directorate S were decimated—namely Department 4, which covered the Americas but also the European division in the wake of 2010—but the directorate's work in other parts of the world seem to have survived Poteyev.

In the summer of 2014, the FBI closed in. As with Ghost Stories, they needed evidence. Buryakov met two people representing wealthy investors who wanted to develop casinos in Russia. Sporyshev warned Buryakov it might be a trap. He was right. One was an undercover FBI source, another an undercover FBI agent. But he ignored the warning. The source invited Buryakov to his office in Atlantic City, New Jersey, and they went on a tour of casinos. The source offered a document labeled "Internal Treasury Use Only," which contained a list of Russians who had been sanctioned by the United States. In a world in which oligarchs and the Kremlin worried about their money, this was a top priority for Moscow. The source said he could get more if Buryakov was interested. Buryakov made clear he was eager for anything on sanctions. On a cold winter's morning in January 2015, Buryakov was buying groceries at the local store and the FBI pounced. He was sentenced to thirty months for failing to register as a foreign agent. Sporyshev and Podobnyy had diplomatic immunity. But one particular encounter with an American would cause waves years later.

ON THE MORNING of Friday, January 18, 2013, a small crowd gathered at 725 Park Avenue, home to the Asia Society. They were enjoying breakfast before a symposium titled "The Triangle of Sino-American Energy Diplomacy." It was the kind of talk that attracted a niche crowd. In the audience was an American who, like others, came to these events to make connections. His name was Carter Page and he had been fascinated with Russia for much of his adult life. It started when he was young and he had come home from skateboarding one day to see Ronald Reagan and Mikhail Gorbachev on TV meeting in Reykjavik to discuss nuclear weapons. He joined the navy and first visited Moscow in June 1991, just before the coup, as a US Naval Academy midshipman. After working as a surface warfare officer, he left to try to make his fortune at the point where business and foreign policy intersected, moving between London, New York, and Moscow. He had lived in Russia between 2004 and 2007, working as deputy manager of Merrill Lynch's office in Moscow, specializing in energy deals and working with the Russian energy giant Gazprom. He then set up a firm, into which he had plowed his life savings, that advised on investment in the energy market. He also received a PhD from the University of London and was linked to various American think tanks and policy institutes, including working as an adjunct professor at NYU. Russia was always his focus and Page says he has always sought to "turn the increasingly dangerous tide" between the United States and Russia.

At the Asia Society symposium, one of those he met was a young attaché from the Russian consulate in New York. He introduced himself as Victor Podobnyy. "He happened to be in the audience, and we struck up a conversation," Page would later explain. The two traded some emails and a couple of months later in March they met again for a Coke.

Page says the Russian never asked him for anything. It was just a discussion about international relations. The two did, though, talk about Gazprom, the Russian energy giant, which Page had advised. He talked to Podobnyy about the shale gas revolution that was taking off in America. Page had a relationship with one of the largest producers and it was looking to increase demand for natural gas. Russia was looking to do the same. There was more contact in June 2013.

Podobnyy did not think that much of Page. "I think he is an idiot and forgot who I am," he complained after Page failed to reply to a message. The Russians discussed how Page had "got hooked on Gazprom"—the giant Russian energy company. "[I]t's obvious he wants to earn lots of money," he said. Podobnyy told Sporyshev that he could use his cover role to dangle contracts in front of Page. This is a classic Russian trick—reel someone in by promising a juicy business deal down the line. But the deals were often a mirage. Podobnyy laughed about it when speaking to Sporyshev and added: "I will feed him empty promises. . . . This is intelligence method to cheat. You promise a favor for a favor. You get the documents from him and tell him to go fuck himself." Page may never have been reeled in, but the case provides an insight into the way the Russians operate. There is no shortage of people who want to make money and see deals with Russia as a way to achieve that. Often the Russians do not need to approach them since these people will approach Russians, who themselves have links to the intelligence services. They can then be studied to determine if they might be useful then or in the future. The possibility of a juicy contract will be enough to get things moving and is an effective way for the Russians to exploit not just the openness but the greed of some in the West.

In June 2013, Page says he was interviewed by FBI agents at the Plaza hotel. He acknowledged he emailed some materials to the Russian but said they were just documents he was using to teach a class at NYU titled "Energy in the World." Page says Podobnyy showed little interest anyway—"his eyes were kind of glazing over." He said he did not know the man was a spy. He also suggested the FBI had better things to do—especially given that a bomb had gone off at the Boston Marathon two months earlier. The FBI never believed that Page was recruited as an agent in 2013 (nor in a previous contact in 2008 with a Russian intelligence officer named Alex Bulatov, who worked at the Russian consulate in New York). He may not have become a recruited agent but Page would illustrate how individuals could end up having influence in ways it was hard to predict when they first came into contact with Moscow's spies. "This isn't about just stealing classified information. This is about stealing you," Maria Ricci said soon after the case became public in 2016. "It's about having you in a Rolodex down the road when they need it." Russia was shaping up for a new spy war—one in which the nature of espionage—and the role of illegals—was changing. It was becoming a battle for influence as much as secrets.

25

A New Conflict

THE NEAR-NAKED RUSSIAN fighter had just pummeled his American opponent to the floor. The crowd had got what they came for. Then Vladimir Putin strode confidently into the ring to congratulate the Russian on his victory in the mixed martial arts bout. It was November 2011 and there were more than twenty thousand fans inside Moscow's Olympic Stadium and millions more watching on live TV. Putin was a martial arts fan. These were his people. But then something strange happened. The crowd seemed to whistle, jeer, and boo Putin. Russia's prime minister struggled to make himself heard as he lauded the Russian "warrior." Officials would later claim the heckling was for the defeated American, but the clip went viral on Russia's vibrant social media scene and was quickly viewed millions of times. Putin had just announced he intended to return to the presidency. And something was going wrong.

A WEEK EARLIER, Putin had met a very different crowd. He'd kept them waiting but they were not the type to jeer. It was the annual meeting of the Valdai Group—experts on Russia from around the world who were given an annual tour of the country and an audience with Russia's leader. This year they were due to hear from him at a restaurant in Kaluga, a suburb of Moscow. It was situated next to a horse-riding school where Russia's leaders kept the most beautiful horses that had been gifted from other countries, particularly Turkmenistan. As they waited for the guest of honor to arrive, they learned the whole restaurant had been redesigned just for that one meal—there were even some endangered fish floating around in a tank to add to the air of opulence. One person leafed through the pages of a property magazine lying around for the rich clientele who were normally there. It listed properties in Britain—the cheapest of which was more than four million pounds. Finally Putin arrived, nearly three hours late. There was no apology. He strode in and raised his arms. An aide walked up and took off his outdoor coat. Another aide slipped a casual coat onto his shoulders. Putin then sat and took questions for more than three hours, until well after midnight. He dwelled on how he had created stability after the chaos of the 1990s. But what those who attended the dinner and the tours before it recall is that despite the show of power there also lurked a sense of staleness and even insecurity. Beneath the surface they could feel a nervousness about the upcoming election and a disillusionment among ordinary Russians they spoke to. Putin and his people were right to be worried.

Another recent event had particularly disturbed Putin. On October 20, Colonel Muammar Gaddafi, the deposed Libyan leader, was cornered by rebels in a large drainage pipe. He was brutally killed, his corpse mutilated. Putin said he was "disgusted" by what had happened. An element of personal fear that this was how things could end lay behind the comment. But events in Libya had also been the cause of profound shock in Moscow and hastened the death of the "reset" with Washington. In spring 2011, President Medvedev abstained on a UN Security Council motion that allowed the West to intervene in Libya after Gaddafi looked set to unleash his military against protesters. Medvedev's move deeply angered Putin, who felt that his protégé was naive when it came to the West. The dynamics between the two had been poorly understood by the West, which had invested heavily in the new leader but had failed to appreciate that the old one was not gone. Medvedev then found himself undercut and feeling betrayed when the West went ahead with regime change in Libya. The West could not be trusted, it seemed to confirm, strengthening Putin's hard-line view.

THREE WEEKS AFTER the martial arts bout, street protests broke out in Moscow. The results of parliamentary elections were viewed as rigged to favor Putin's party. These protests did not involve the majority of the country; rather it was primarily liberal, urban, younger people who had thought their country had moved on but were now told they were instead going back to Putin. The protests humiliated the Russian leader but also scared him.

Putin's fears of regime change had been heightened by events in Libya and indeed across the Middle East and North Africa, where the "Arab Spring" was under way. Mass protests seemed capable of toppling corrupt, decrepit regimes. In March 2012, Putin would win his election comfortably, taking him back to the presidency by stressing he would keep the country strong and stable and avoid a return to the 1990s. But the events of the previous year had left a deep mark. The Putin who returned as president was a different Putin from the one who had left. In his first period as president he had uneasily straddled the hard-line forces, suspicious of the West, and those who felt Russia needed to modernize. Now he sided with the former. It took too long for the West to understand this change.

Putin and his allies blamed a hidden hand for guiding the protests in Moscow and the Middle East—and that was America and the CIA. This was the familiar belief—going back to Soviet days—that US intelligence was interfering and subverting. It was also convenient to blame protests on outside support rather than acknowledge genuine domestic disillusionment. But many in Russia's intelligence services, like Putin himself, had a conspiratorial worldview, which meant they genuinely believed this theory. US secretary of state Hillary Clinton's remarks during the December 2011 Moscow protests touched a nerve. "We are supportive of the rights and aspirations of the Russian people to be able to make progress and realize a better future for themselves," she said, a set of comments probably seen as doing no more than reflecting the disappointment in Washington at Putin's planned return. But Putin interpreted this as a "signal" to protesters. "We all understand the organizers are acting according to a well-known scenario," he noted. The same tricks Washington had used in Ukraine and Georgia and the "Color Revolutions" were being used to meddle in Russia's domestic political affairs, he believed. Putin would have his revenge for that. And he would do it through his own judo slam, turning his opponents' strength against them.

The West was working through domestic puppets, in Putin's eyes, like Russian NGOs. Mike McFaul had arrived as the new US ambassador in December 2011. Since he had long-established contacts with democracy promotion groups, his appointment was immediately portrayed as part of the master plan to foment revolution. McFaul, one of the architects of the reset, would become the victim of its failure as a campaign of harassment against him intensified. He initially thought the claims of American subversion were just talk. But by the end of his difficult time as ambassador, he would realize Putin genuinely believed the United States was engaged in trying to subvert his grip on power. There is something like a "deep state" in Russia, largely made up of current and former spies, and Putin believed the equivalent group in America was determined to maintain conflict with Moscow. Putin is said to enjoy watching *House of Cards,* a drama about devious manipulation in American politics. He is also said to recommend it to people in order to understand America, believing it is an accurate reflection of how Washington works. Drama and fiction—especially spy fiction—often shape how countries think of how the world really works and what their adversaries are up to. That has certainly been the case with the West and Russia.

SPY FEVER RESURFACED in Moscow. In May 2013, an FSB officer pulled the baseball cap off a dejected American diplomat in front of TV cameras to reveal a poor-quality blond wig underneath. The FSB said the third secretary at the US embassy had been caught carrying a map, compass, cash, and a letter offering $100,000 up front and $1 million a year to spy for America. The recipient was told to go to a cafe and open a new Gmail account and send an email to a specified address. His target was alleged to be a member of the Russian security service who worked on the North Caucasus, whom he perhaps was meeting under the cover of liaison discussions about counterterrorism.

Russia's enemies now had a new weapon alongside spies, the Kremlin believed, and that was the internet. When Putin took over at the end of the 1990s, his focus had been on bringing mainstream media like TV channels under tighter state control. The internet was largely left alone. Bloggers like Alexei Navalny had made their names exposing corruption. But 2011 had been the turning point. Popular uprisings that swept away authoritarian regimes in Tunisia and Egypt were described (only partly accurately) as "Twitter and Facebook revolutions." Silicon Valley and its friends claimed the new tools were allowing ordinary people to communicate and challenge power. The Kremlin realized this could be a threat and began to exert more control over the "information space." There was more surveillance and filtering of internet traffic. Social media companies—domestic and foreign—

were put under more pressure, as TV had been before. Putin would make clear his thinking with a telling remark at a public forum. The internet had been originally developed as a "CIA project" and "special services" remained at the center of it, he said. The internet, he believed, was a tool of subversion to spread "Western" values.

Putin's case was aided by a surprise arrival at Moscow airport in 2013. Former NSA contractor Edward Snowden was on the run, having disclosed details of some of the surveillance agency's most secret programs. Escaping Hong Kong, he found himself stranded in Moscow. In Washington, the Obama administration was frustrated with Moscow's sheltering of a fugitive whom it wanted back much more than it publicly let on (it forced the president of Bolivia's plane to land because of suspicions he might be smuggling Snowden out of Russia). In discussions in the White House, the CIA cautioned that the chances of Russia handing Snowden over were low—partly as that would risk making Moscow look like a risky place to go if you were a defector. There is no evidence that Snowden wanted to stay in Moscow or had been a Russian agent. He was motivated by his own libertarian ideology. But Putin knew how to take advantage of an opportunity that had fallen into his lap. Snowden exposed the vast power of America's NSA and Britain's GCHQ to tap global communications as well as the apparent complicity of American technology firms that handed over user data through programs like PRISM (on the basis of secret legal orders). The disclosures were ammunition to the claim that the "free global internet" was actually a Western surveillance machine. A dependence on Western technology led to vulnerability. The answer was to assert domestic control. The Snowden revelations also caused real damage to America's relationship with some of its closest allies. That was because the NSA was revealed to have been spying on friends as well as enemies. One particularly damaging story was that the NSA had been listening to German chancellor Angela Merkel's cell phone. There was something curious about that story. It was one of the only revelations at the time that had details of a particular target rather than the PowerPoint presentations and the like that Snowden downloaded. Senior US intelligence officials from the time had a theory that this revelation came not from Snowden but the Russians. Possibly by some other intelligence-gathering operations, they had gathered the information and then pushed it out during the Snowden deluge of stories as a so-called active measure to damage US-German relations. If so, it worked. The central front in the conflict between the West and Russia was changing—away from military competition and even traditional espionage to something new.

IN FEBRUARY 2013, a military newspaper published a speech by the Russian chief of the General Staff, Valery Gerasimov. A new era of warfare had dawned, he explained. It involved "political, economic, informational, humanitarian and other non-military measures"—supplemented by "covert military measures," including the use of information. The article would lead to discussion, sometimes misguided, about a "Gerasimov doctrine" and hybrid warfare. What was often missed was that it was written not as a template for Russian warfare but as an explanation of what Russia thought it was being subjected to already by the West. Russia was a besieged fortress—and it needed to learn how to fight back.

The distinction between war and peace was becoming blurred. Conflicts would no longer begin with a declaration of war followed by tanks rumbling across Europe. Victory in "hybrid warfare" would belong to those who could use nontraditional means to achieve their objectives. Information was a weapon that could be deployed to destabilize a country short of traditional force. And so Russia began to mobilize. Some of this had nothing to do with espionage and covert action. In October 2014 it was announced that Russia Today (rebranded RT) would launch a dedicated channel in the United Kingdom. That summer Putin had said the aim of the network had been "to try to break the Anglo-Saxon monopoly on the global information streams."

When protests began in Ukraine in February 2014, Russia attributed them to Facebook and Western covert influence. The West failed to understand how far Putin would go to prevent Ukraine from being pulled out of Russia's orbit and into that of the EU. Putin improvised rapidly and intervened. In the crisis—and especially the seizure of Crimea—the Kremlin was able to implement and road test elements of its new doctrine of hybrid warfare. Information and propaganda campaigns were launched, along with covert action. In some cases, action was not very covert. Men in uniform began to appear in Crimea who were christened "Little Green Men." The Russians denied they were their troops when it was obvious they were. It was an example of a new form of brazenness.

Putin was right in thinking the West would not go to war, but he failed to see it would respond with sanctions that isolated Russia. That fueled Putin's nationalist agenda and allowed him to pressure oligarchs to decide whether their priority was their houses in London or their loyalty to him. In Ukraine and also in his intervention in Syria to save his ally President Assad, Putin was succeeding in making sure Russia could not be ignored. But at a price for his own country.

GETTING INSIDE THE mind of your adversary is a key ambition of spies. In Putin's case, senior British spies say it is particularly hard since there are very few decision makers around him who matter (although the CIA may have been able to get at least one source close to Putin through one of his advisers). Russian expertise had atrophied in Washington and London since the end of the Cold War. But the same was also true for Putin. He had an enormous intelligence apparatus but the man at its center had very little feel for how the West really worked. His mind-set remained that of the young KGB officer in Dresden. The goal of Putin was to ensure stability at home and in the near abroad, and deal with those trying to undermine it. Because he believed the West was trying to undermine his grip, he believed it had become vital to keep the West off balance and divide it.

And so by 2014 Russia saw itself engaged in an ongoing conflict with the West—one below the threshold of war and fought with spies, hackers, and information. The West still thought of a simple distinction between war and peace and was slow to understand that things were changing. One side was engaged in this conflict—the other had not even realized it had begun. "We fail to appreciate their doctrine—partly as we don't want to face up to what that means," says one of Britain's most senior spies.

IN THE PERIOD after the Ghost Stories investigation, Russia remained fuzzy and out of focus in the eyes of Western policy makers. Counterterrorism was still the overriding priority for the FBI. Even within counterintelligence, resources were shifting toward China. Russia specialists warned this might be a mistake. The view of Russian espionage in Washington and London was that Moscow had a traditional way of doing things—always had and always would. But from 2012, Moscow was innovating in a way that was being missed. In the CIA, one of those who worked in Russia House who had been skeptical of the "reset bullshit" remembers the period from 2012 as a bit like the late 1990s when it came to the threat from Al Qaeda and terrorism. The warning lights were blinking red. But no one outside their small world was listening. "What did we do about it?" one person deep in Russia operations during those years asks of US policy. "Zero. Nothing." There was a failure of imagination, just as there had been with terrorism. One place where the warning signs were most clear was online.

CYBERSPACE IS THE natural home to hybrid warfare. It allows just enough remoteness and deniability—what some call implausible deniability—to make it easier for one country to act against another without moving to full-scale conflict. And this was one area where the Russian spies were highly capable. Cyberspace would prove to be one of the key ways in which Russia would reinvigorate its intelligence capabilities and push them in new directions.

The end of the Cold War had coincided with the internet's spread (almost to the day, since the World Wide Web had been released to the general public the same month as the 1991 coup in the USSR). America was the dominant player. But Russia had niche skills—it had always prided itself on its mathematical training—and a new generation who came of age in the chaotic Russia of the 1990s would find a home in the darker regions of the wild west of the Web. There is a narrative that links Russian cyber activity primarily with criminal hackers. The story is that hackers were recruited by organized crime gangs who were told by the state they could steal what they wanted abroad as long as they did not cause trouble at home. A bit like the oligarchs, they were tolerated so long as they understood their place and occasionally did a favor when ordered. This is true. But it overlooks another story. The real high-end Russian cyber espionage has always been carried out not by criminal hackers but by Moscow's spy agencies.

In 1996, a young US Air Force cyber expert named Kevin Mandia was asked to respond to what turned out to be the first cyber espionage campaign to steal American military secrets. It was code-named "Moonlight Maze." "Everything went back to Moscow," Mandia told me years later. "They spoke Russian. They hacked out of Moscow." It had been carried out by hackers within the FSB (which eventually absorbed much of the KGB's signals intelligence capabilities). That same group—code-named "Turla"—would breach the US military in a serious incident in 2008 code-named "Buckshot Yankee," when a soldier inserted an infected USB left by the Russians in a parking lot abroad. The code worked its way deep into classified systems and took more than a year to clear out. This group remains the most advanced, stealthy cyber-espionage threat actor, one that often eludes those trying to track its movements. Its aim for many years was simple—steal classified information.

This was the traditional world of stealing secrets, but Russia was also learning how to employ digital techniques to have real-world effects. In Estonia in 2007 the Baltic country's institutions were subjected to massive denial-of-service attacks after a row with Russia. In Georgia the following year, Russia combined cyber and information operations with regular warfare. But it was from 2014 that Russia pushed the boundaries in Ukraine, hacking and then leaking information. Surveillance technology picked up an American diplomat's angry private comments about Europeans, which were spread on social media. The Russians even took a power station off-line for a few hours in a carefully controlled act to test destructive capability. Ukraine was a testing ground but Russia was also beginning to deploy its more aggressive tactics farther afield.

MANDIA, NOW RUNNING a large private-sector cyber defense company, got a call from one of his "first responders" in mid-August 2014. They were responding to a breach at a government agency and they thought their boss would be interested. Mandia had been watching Russian government hackers back to the mid-1990s.

"The rules of engagement never changed. So you're talking about a twenty-year run where it's the same behaviors," he says. But that August something was happening. APT 28 (APT standing for Advanced Persistent Threat)—the GRU—was becoming more aggressive and looser. He even started to see APT 29—the normally stealthy SVR hackers—getting noisier. Usually, when these hackers realized they were being observed, they disappeared and went quiet. But now they were not bothering to hide their tracks. That summer he also saw several universities compromised with the aim of stealing the emails of professors critical to Putin. That strayed from the traditional targeting he had seen before. "Suddenly, their rules of engagement had changed," Mandia says. "It was another warning sign. The unraveling of a known situation." Those who had watched closely over an extended period, whether in Russia House of the CIA or like Mandia in the cyber world, could tell that something was changing. The NSA maintained what one official at the time calls a "caretaker" capacity on Russia. Despite the agency's huge size, this was a tiny group that acted as a tripwire. If they saw something significant, they could ring an alarm bell. From 2014 onward, the small group was ringing the bell as loud as it could, the former official says, but no one acted on that knowledge.

In February 2015, Mandia's team saw some strange traffic coming out of a French TV network—it seemed to be linked to APT 28. Little did they know that this would result two months later in Europe getting a major wake-up call.

APRIL 8 WAS a big day for the French TV network TV5Monde. The network had just launched a new international channel. Ministers had been in attendance. There was a party at a fancy restaurant in Paris that evening to celebrate. At 8:40 p.m., just as Yves Bigot, the director general of the network, was being served his appetizers, his phone lit up with calls and texts. He excused himself from his guests. He was given some bewildering news from the office. All twelve of the network's channels had gone black. They had been taken off the air.

The technical team from the channel battled to work out what was going on. As each second passed they could see more critical systems were being corrupted. No one could understand why. "We were a couple of hours from having the whole station gone for good," Bigot later told me in his smart Paris office. Fortunately, the launch of the new channel meant their engineers were on hand and one was able to cut off the connection through which the attack was coming from the internet. The phrase "cyber attack" is often used loosely for some minor breach or attempted breach of a system. But in this case it really had been an attack. And it soon became clear it was a highly sophisticated one. The attackers had first penetrated the channel months earlier to carry out reconnaissance. They had used multiple points of entry, even going through the Netherlands-based system used to remotely control the TV cameras in the studio. They had then designed bespoke computer code to destroy the hardware that controlled the TV station's operations before unleashing it on April 8. The result was devastating. The company had to return to fax machines and wait months before it could reconnect to the internet. The cost was in the millions of euros. But one British intelligence source says the strike could have been even worse, if those behind it had wanted that. But who was behind it and why?

At the same moment that Yves Bigot received the call in the restaurant on April 8, messages had been posted on the channel's Twitter and Facebook pages. They said the hackers were from a group calling themselves the "Cyber Caliphate"—masked jihadists appeared to be making threats against France. French investigators contacted counterparts from Britain's GCHQ to help. As they looked into the attack, they got a surprise. The Cyber Caliphate was what is known in intelligence as a "false flag." The network had actually been attacked by a group of Russian hackers from the GRU's APT 28. What the GRU had done to TV5 was not espionage—its premeditated aim was destruction. This was something new.

All of Russia's intelligence agencies knew what the small group in the Kremlin wanted to hear and understood it was in their interests to provide it rather than challenge it. The SVR offered to explain the world. But—not least because of the 2010 embarrassment—it was falling out of favor. Its rivals, meanwhile—the GRU and the FSB—offered to change the world. The GRU was deeply involved in Ukraine and also Syria, and if Russia was going to fight information warfare, then that meant that the GRU would muscle its way to the front of the pack. It had the ability to blend the real and the digital, carrying out cyber operations and traditional media campaigns in support of semi-covert military activity, and now it was clear it would even carry out destructive cyber attacks.

In London, the Joint Intelligence Committee, composed of senior intelligence and national security officials, struggled at their meeting to understand why TV5Monde had been targeted by the GRU (the CIA was reluctant at first to believe it was even possible). The conclusion was that it was likely an attempt to test cyber weapons under a vague cloak of deniability. "The best-case scenario was that the hackers were out of control; the worst case was that they were under the control of the state," one spy chief thought. The concern was that a British general election was due just weeks after TV5 was taken down. What if British TV channels were taken off air on election night as the results were coming in? What would that do to trust in the democratic process? Broadcasters were urgently contacted to warn them of the risk and told to draw up contingency plans.

GCHQ divides up the world and its targets with NSA. The British agency had the lead on watching the GRU. The spy agency used its global signals intelligence capability—the ability to intercept traffic flowing through data pipes—to watch GRU activity. It effectively spied on a group of burglars and vandals as they went around different neighborhoods causing trouble. It often saw "indicators of compromise" of data being stolen and then sent back via a circuitous route to Moscow. In April 2016, GCHQ informed their American allies that they saw the Russians breaking and entering one of their properties. Nearly half a century earlier, it had been real-life burglars; now it was Russian cyber thieves who had broken into the Democratic National Committee. But as the United States entered an election year, it was not the only way that Russian spies—including a new type of illegals—were at work as a new conflict gathered pace.

26

The New Illegals

THE WHITE PICKUP truck driving around West Palm Beach, Florida, in August 2016 had a couple of American flags attached to the rear but also something you saw less often. On the back was a steel cage. Inside were some cheap plastic chairs and seated on them were two people. One was a pretty poor imitation of Bill Clinton, not nearly jowly or portly enough. Sitting next to him was a woman in a second-rate Hillary Clinton mask, in a T-shirt with what was supposed to be prison numbers on it. "I'm NOT with crooked Hillary," read the sign on the back. The man who had built the truck had done so after a few phone calls from someone who did not speak great English but offered to pay for the work. What the man building the cage did not know—and like many others organizing rallies across America—was that he was being manipulated by a new breed of Russian illegals.

These new illegals were everywhere in America and nowhere. They were deep in communities but impossible to see. They were cajoling and criticizing, coordinating and controlling, and yet always out of reach. They were the new ghosts who, for too long, the authorities did not even know existed. And it was in the election of 2016 that they first made their presence felt.

• • •

RUSSIA SPYING ON American politics was not new. Back in the 1970s, the KGB had recruited a Democratic Party activist who reported on Jimmy Carter's campaign and foreign policy plans. The Ghost Stories illegals were tasked with collecting intelligence on politics and the 2008 election. And Russian intelligence in the run-up to the 2016 election carried out intelligence collection against campaigns, think tanks, and lobbying groups. But this was all classic espionage of the type seen for decades—understanding who was up and who was down and what policies they might pursue in power that would affect Russian interests. This time it was going to be different. As well as espionage, there would be subversion. This should not have been a surprise to those who had watched Russia closely since the end of the Cold War. Putin and other former KGB officers around him blamed that defeat on subversion by the West and believed political interference by Western intelligence had never stopped since. Now they were going to respond. And there were new weapons in their armory.

THE LAKHTA CENTER is an eighty-seven-story skyscraper in St. Petersburg, due to be headquarters for the energy giant Gazprom. Project Lakhta, though, was run out of a far more anonymous, squat, modern building in the city. Project Lakhta was the cover name for the Kremlin's overarching operation to influence politics. The squat building in St. Petersburg was home to what became its most famous vehicle for the project—the Internet Research Agency (IRA). There are other organizations with similar innocuous-sounding names, but the IRA is the one that has gained most attention, thanks to a detailed US indictment. It was registered in July 2013 (although some believe it was active earlier). The annual budget would run to roughly a million dollars a month and hundreds of staff were engaged in its work. Its aim was not just to understand what people thought but to change how they acted using modern social media. It grew out of the new Putin regime, scarred by the 2011 protests and determined to manage public opinion. The way to do that was pushing your message out and battling those who were against you. This could be done covertly by using the internet's anonymity to pose as someone who you were not. In its early days, it was seen as simply a "troll factory"—but it would prove to be much more. It was actually something more akin to a factory manufacturing identities and mass-producing new illegals—the cyber variety.

Project Lakhta's initial target was close to home. The Kremlin saw itself as engaged in political warfare in which the West was trying to stir up protest within Russia and its neighborhood. Project Lakhta's job was to combat this. The Kremlin believed that if you control the information space, then you have won half the war. A rash of Twitter accounts first emerged in Russian supporting the government narrative. But these new tools were then turned abroad, particularly from 2014—the year of the Ukraine crisis. English-language accounts appeared as the Kremlin sought to build support at home, undermine Ukraine's unity, and reduce the chance of the West responding to Russian actions. Those inside the IRA considered the expanding "foreign desks" to be more sophisticated than the domestic. Western intelligence officials say the IRA's strongest language capabilities are in Russian and then English. The playbook Russia had road-tested over Ukraine—a mix of propaganda and cyber hacking—would be extended to the Baltics and Poland, on through the rest of Europe and all the way to the United States.

In April 2014, the IRA created a department that went by various names—one was the "translator project." This would focus specifically on the US population and worked on social media platforms like YouTube, Facebook, Instagram, and Twitter. It would eventually have more than eighty employees assigned to it. The month after it was formed, the new team was discussing how to interfere with the 2016 election—still two and a half years away. This was a patient, long-term effort, not a last-minute rush job. Internally it acknowledged it was involved in "information warfare against the United States of America" using fictitious US personas on social media platforms—Russia's new illegals.

To divide America, you needed to understand it. And there was only so much you could do from afar. So one of the first priorities was to carry out field research. Four staff members applied for visas in 2014. In the end only two traveled, arriving on June 4 and staying for twenty-two days, claiming they were friends who had met at a party. But it is alleged they were on an intelligence-gathering operation. This might not have been spying in the way most understood it, but they still took some precautions, including having "evacuation scenarios" if they were found out. It was quite the road trip, including stops in Nevada, California, New Mexico, Colorado, Illinois, Michigan, Louisiana, Texas, and New York. On their return the pair exchanged intelligence reports on the trip. Another Russian went to Atlanta for four days in November. Among those who traveled was the specialist who oversaw the project's data analysis group—a sign that this involved a different skill set from the spies of the past.

This was the evolution of the work that the old illegals had once carried out. The old illegals had the mission of spotting individuals of influence and of understanding where and how power flowed in Washington, New York, and Silicon Valley. But now the target was not the powerful but the common people. If you want to manipulate someone you need to understand them. These new spies were going to "Middle America" to understand what made ordinary people tick—what made them angry, how they talked, how they communicated. This was what you needed to understand, not how power flowed among the elite but how the people felt about the powerful. New tools had made it possible to conduct influence operations on the mass of the population remotely, but there was still the need for a new type of human intelligence obtained on the ground in order to target your work.

This was an updating of an old KGB concept—what are called "active measures." Active measures, or influence operations, were a core part of KGB doctrine throughout the Cold War. The CIA estimated the USSR spent $4 billion a year on active measures in the 1980s, which included trying to influence people through fake stories. The aim was to weaken Moscow's enemies by increasing distrust of the United States or undermining faith in the NATO alliance. The KGB tried to stir up conspiracy theories on John F. Kennedy's assassination and smeared politicians and officials it did not like. It had also tried to influence politics, supporting social movements that might be sympathetic to its aims. In 1983, KGB agents in America were ordered "to acquire contacts on the staff of all possible presidential candidates and in both party headquarters . . . [and] to popularize the slogan 'Reagan Means War!' " Many of the influence operations were abject failures but others—like claims that the US military was responsible for the AIDS virus—spread insidiously. After being first planted in Indian newspapers, it spread over years through sympathetic media in other countries. A new online world of social media offered the chance to do this at scale and at speed.

FOR DECADES RUSSIA feared its dependence on Western technology made it vulnerable. But now it realized it could exploit the vulnerabilities that a connected world created for its Western adversaries. Social media networks—which tech companies promised would help democratize information and empower ordinary people to speak their voice—turn out to be excellent at distributing propaganda, misinformation, and fake news. It was another way in which the West would be complicit in aiding Russian influence. The essence of social media—its speed, its anonymity, its openness, its love of controversy—made it ideal for Russian influence operations. Its decentralized nature allows hidden actors to reach large audiences quickly. Its algorithms can be gamed by those who understand it. Its frictionless anonymity was perfect for the type of identity transfer that spies used to engage in but were finding harder to do in the world of biometrics and databases. The new online world was perfect for a Putin judo slam.

Turning up for shifts like postmodern factory workers at their office in St. Petersburg were IRA staff referred to as "specialists." Their job was to create social media identities that made it look like they were Americans. They were the online version of the storytellers of Department 2 of Directorate S. But if Department 2 were artisans, creating careful forgeries that could withstand expert scrutiny, the new "specialists" were playing cheap and dirty. Anonymity on the internet made their life much easier. As the old joke goes, on the internet no one knows you are a dog. No one knows you are a Russian, either.

The specialists were divided into day-shift and night-shift teams linked to US time zones. The aim was not to inform but to divide. They were given extensive and precise guidance on how to do this. "Our goal wasn't to turn the Americans toward Russia," one worker later told a Russian news outlet. "Our task was to set Americans against their own government: to provoke unrest and discontent."

Young men and women—often using attractive pictures and posing as American—started posting about American politics, particularly expressing comments critical of the administration. Twitter was where it began but the first fake American Facebook accounts were created as early as the summer of 2014, and YouTube and Instagram would also be part of the campaign. They used Virtual Private Networks to hide the fact they were really in Russia—the digital version of a false mailing address or identification. When an account was shut down by Facebook for being suspicious, they would email administrators back and cite free speech and the US Constitution, claiming they were real Americans. After someone spoke to the news media about the IRA's existence in 2015, security was tightened. Staff were told to maintain secrecy but also that they should be proud of their work: "Because every country has their own kind of organization that defends their national interests and distributes civil unrest," managers told them. "This is information war, and it's official."

The fake profiles began to multiply. They were designed to address divisive issues and the growing polarization of American politics offered fertile ground for such an approach. The Russians did not create the divisions in American society, but they could play on them. The team created a page on immigration called "Secured Borders"; one on Black Lives Matter called "Blacktivist"; one on religion called "United Muslims of America"; and another called "Army of Jesus." By 2016, the size of some of these groups had grown to hundreds of thousands of online followers. On Twitter, fake accounts included "Tennessee GOP," which claimed to be controlled by the state party. It attracted more than 100,000 online followers. Automated accounts—so-called bots—would then amplify the messages from the fake American accounts. US media outlets sometimes quoted these Russian-controlled accounts, believing they reflected the views of real Americans.

In St. Petersburg, the "specialists" would receive regular feedback to ensure they looked as American as possible. In the early days it was recommended staff watch *House of Cards* to understand how American politics worked, but by the latter stages they had evolved significantly in their subtlety and were providing tutorials on the American tax system so that the posts would hit the right note. Their advantage was that they were flexible, always experimenting, seeing what might work, and discarding things that did not. Activity would spike at the time of political events like candidate debates in the primary or general election.

US intelligence believes influence campaigns were approved at the highest levels of the Russian government—particularly those campaigns that were politically sensitive. This operation was not run by the SVR, GRU, FSB, or any other intelligence agency, though. It was, according to a US indictment, allegedly funded by Concord, a company owned by businessman Evgeny Prigozhin. He was often known as "Putin's chef" and his company had contracts with the government to feed schoolchildren and the military. He and the company have strongly denied the allegations against them. This was another sign of how Russian activity had changed—the Kremlin was now working through a wider range of proxies than just its spies.

As early as 2014, there had been scraps of intelligence that Russia was looking at how social media could influence elections. That year the Obama administration reportedly received an intelligence report suggesting that the Kremlin was building a massive machinery of disinformation that could be used to interfere in politics, according to a later report. "You have no idea how extensive these networks are in Europe . . . and in the US, Russia has penetrated media organizations, lobbying firms, political parties, governments and militaries in all of these places," a Russian source is reported to have said. But no one quite understood the risks at that point. This was despite the fact that some of what Russia was capable of was being witnessed in front of people's eyes in Ukraine, but also, to a lesser extent, the Baltics. The Russians were starting to try out their new weapon in hybrid and information warfare in its near abroad. "We did not think they would have the balls to see it in the US or UK," acknowledges one senior counterintelligence official from the time. Former FBI officials from the time acknowledge they were slow to recognize the evolution of Russian tradecraft and especially the shift to use technology and social media. It is only in hindsight, they say, that you can see the trajectory. It was only in October 2016 that the FBI's counterintelligence division asked a contractor to look for signs of Russian influence on Twitter. There also seemed to be a reluctance to go back to Russia in the upper echelons of the Obama administration after the failure of the "reset." The assumption was that Moscow was an irritant and no more. Certainly not a strategic threat. As before, the Western gaze wandered while that of Moscow remained steady. Meeting only limited resistance, the Russians kept pushing.

THE NEW POLITICAL warfare (a term coined by the United States in the early Cold War for what it was doing in Europe to combat communism) was partly played out on social media but attempts to disrupt the West took place in the real world as well. If the West was divided and off-balance, Russia had more room to maneuver to pursue its own interests. The Kremlin began to step up its engagement with extremist groups left and right across Europe. Sometimes they would be pro-Russian, but often they were simply disruptive of the pro-EU and pro-NATO political consensus. On the right, this fit with Putin's move to a more nationalist, orthodox worldview that focused on protecting traditional values against their erosion by Western liberalism. On the left, it focused more on anti-Americanism. Oligarchs sometimes paid for flights of European politicians to go to Russia or offered them business opportunities. Russia has invested heavily in long-term influence agents, the head of the Estonian intelligence service has said. It looked for people on the margins of political life and offered them money and media support. These may well be people either spotted by old illegals years ago or run by them now. Not all of these investments paid off but enough did to ensure there are people in national parliaments pushing a pro-Russian agenda. Such influence—as opposed to traditional espionage—exists in a gray area and can be hard to define and to challenge.

IT WAS ONLY looking back after 2016 that US intelligence officials would understand the extent of Russian ambition and put all the pieces together. Despite the presence of sources in Moscow, Western spies lacked the appreciation of Russian intentions, especially since operations were not always run out of the traditional institutions that were the targets of Western espionage. Spy-catchers talk of "collection bias," in which they focus on things that are secret or people stealing secrets, but this often leads to them not spotting influence operations. The dilemma posed by the 2010 illegals foreshadowed what was to come. If your opponent was not stealing secrets, were they really an intelligence threat? The point about Russian influence operations was that they were not about "secrets"—they were about influence. This was how Russian hybrid warfare worked. The wariness of spy-catchers getting involved in politics also made them sometimes reluctant to look too closely into issues like party funding or contacts of politicians. The result was that almost no one was watching the new Russian illegals as they went about their work. Russian interference in the 2016 US election would be a successful intelligence operation (at least in the short term), yet it had nothing to do with classified material.

AS THE ELECTION year began, the Russians sharpened their focus. February 9, 2016, saw the New Hampshire primary. The next day in St. Petersburg, a plan was circulated. Specialists were instructed to "use any opportunity to criticize Hillary and the rest (except Sanders and Trump—we support them)." Putin held a grudge against Clinton for what he perceived as her role encouraging the 2011 protests. It was personal. And now it was time for payback. Former workers say the IRA was targeting her a good two years before the 2016 election.

The new illegals did not just influence the online world. They could make things happen in the real world by organizing rallies and counterprotests, engaging real Americans by pretending to be US grassroots activists and getting them to do things. In some cases—in a high-tech echo of the doubles of the past—the Russians stole real people's identities. They gathered Social Security numbers, home addresses, and dates of birth of real Americans. They then used those to open PayPal accounts to pay for ads on social media sites or for expenses for political rallies. Other times, communications were made and ads paid for by entirely fictitious US personas created by the IRA on social media. Whatever the case, the identities of these illegals could be built in hours, not years, like that of Heathfield. Of course they would not stand up to the same level of scrutiny. But they did not need to because they were disposable and they were legion.

MATT SKIBER WAS one of the new illegals. He got involved in politics in May 2016 even though he did not exist—a ghost controlled by a person in a computer in St. Petersburg. His Facebook account contacted a real American to ask them to act as a recruiter for a "March for Trump" rally in New York the next month. Ads were purchased on Facebook and Skiber contacted a real person to give them money to print posters and get a megaphone. Next, on August 2, Skiber sent a private message on Facebook to a real account, "Florida for Trump": "What about organizing a YUGE pro-Trump flash mob in every Florida town?" On August 18, the real "Florida for Trump" Facebook account responded with instructions to contact a member of the campaign. In August, Skiber recruited a real American over Facebook to acquire a prison costume. Another person was paid to build the cage for the pickup truck. A Twitter account then organized for an actress to dress up as a caged Hillary Clinton in prison uniform in West Palm Beach. By August 24, more than one hundred real Americans had been contacted through fake US person accounts. None knew they were dealing with a Russian—just like those who had met Donald Heathfield or Cynthia Murphy. The IRA carefully tracked the real people—listing their contacts, political views, and the activities they had been asked to perform. They had unwittingly been recruited into a Russian intelligence operation. They came from across the political spectrum and included black social justice activists as well as moderators of conservative social media groups. Hundreds of real-world rallies were organized by the Russians remotely over social media. They would ask followers on social media to attend. Out of those who expressed an interest, they would pick an American whom they could ask to take over organizing the event, finding an excuse for why the original organizer could not make it on the day. They would promote these rallies in advance by contacting US media and putting them in touch with the real American coordinator and publicize the events afterward through their own Russian-controlled social media accounts.

A cyber-hacking campaign had gathered pace at the same time. On March 19, Clinton campaign chairman John Podesta clicked on a link that looked like a security notification from Google telling him to change his password, but which instead gave GRU hackers access to fifty thousand emails. This was the result of a targeted reconnaissance campaign. The next month the GRU was inside the DNC and Democratic Congressional Campaign Committee (DCCC). They were able to take screen shots of employees' screens and began carrying out searches. In May, the DNC asked a cyber security company to investigate. They found not only the GRU but also a second Russian espionage team from the SVR who had been there back at least to July 2015. The GRU's noisier approach had blown the cover for its stealthier rivals. "We did attribution back to the Russian government," Shawn Henry, chief security officer of CrowdStrike and a former executive assistant director of the FBI, told me soon after. "We believed it was the Russian government involved in an espionage campaign—essentially collecting intelligence against candidates for the US presidency." The DNC went public. But then something happened that no one had predicted. The SVR had been carrying out the type of espionage of political campaigns that had been going on for years but the GRU had been planning something else.

This was going to be an influence operation, an "active measure"—what became known as "hack and leak." With the help of another GRU unit (based at "the Tower"), some of the material was published on a front website and other emails were allegedly transferred to the WikiLeaks organization. There was also contact with journalists, offering them juicy exclusives. On July 22, three days before the Democratic National Convention, damaging material appeared suggesting a bias in favor of Hillary Clinton and against Bernie Sanders, leading to the resignation of the DNC chair. It does not take a lot of work by the hackers to analyze the data. Instead, in another piece of Russian judo, they used the competitive and free Western press to do the leg work of finding stories amid a mass of data that was published and, as a result, carried more weight.

A final component of the cyber campaign within the United States was to undermine the credibility of the election. The first signs of Russian intelligence researching US electoral processes and technology went back to 2014. GRU hackers in June 2016 researched state election boards looking for vulnerabilities. In

August they hacked into computers of a company that supplied software used to verify voter registration information and sent more than one hundred emails to people involved in elections in Florida counties to try to get into their accounts. The likelihood—also from evidence of their social media plans—was that they were going to claim voter fraud after the expected Hillary win in order to delegitimize her election.

This was a multifaceted campaign but the cyber and social media aspects were not always coordinated. That was the way Russia worked—lots of actors were each trying to compete and show off to the Kremlin what they could do to serve their masters and pursue an overall goal set from the top. Covert activity—using false identities—was blended with overt information through Russian media outlets like RT. Too often those in the West focused on one element of this activity—hacking or social media—but failed to see the full spread. Intelligence had been coming into the Obama administration but it had not been pieced together. National security officials were convinced Clinton would win and feared that if they called out Russia for its campaign against her, they would be seen as trying to assist her. A response could wait until after election day.

There were other warning signs that the rules had changed. On June 6, a yellow taxi pulled up outside the US embassy in Moscow late at night. CCTV shows a man striding toward the main entrance. But as he does, a Russian runs out of the sentry booth outside and violently body-slams him, first into a brick wall and then into the ground. The two begin wrestling, the American underneath desperately trying to kick off his assailant. He somehow manages to slide, with the Russian still on top and punching him, toward the main door, which opens. He kicks first one leg and then the other over the threshold and into the embassy and onto American soil. The Russians said the American was a CIA officer.

The attack was more serious than the arrest or harassment of diplomats under previous bouts of spy fever.

A few weeks later, in July, Carter Page, who had come up in the FBI's Buryakov investigation, was in Moscow. He was there to give a speech on improving relations. He says he went as a private citizen, but there was considerable interest in him because in March he had been named as an adviser to Donald Trump's campaign. He says he saw in Trump a man who might fix the problems that had damaged US national security since the end of the Cold War. He had promoted himself as a man who had good contacts, emailing the campaign in January saying he had spent the past week in Europe and had been in discussions with individuals who had close ties to the Kremlin and who recognized Trump could have a "game changing effect . . . in bringing the end of the new Cold War." He said his contacts might be able to set up a meeting with Putin and he also criticized sanctions on Russia. During his Moscow visit in the summer he talked to an old Gazprom contact who now worked at Rosneft, the oil company. There may have been some passing remark by the Russian about the sale of a stake in the company. There were also other conversations about possible business deals. The speech attracted negative attention. Soon Page was no longer part of the campaign, although he would try to get an administration job. What he did not know until later was that he had been subject to a wiretap by the FBI, which would become highly controversial amid accusations that a "deep state" had been "spying" on one of the candidates.

AT THE SAME time, a former MI6 officer was investigating Russian activity and had come across Page's name as well as others. As a young recruit, Chris Steele had been one of the handful of MI6 spies operating in Moscow at the time of the August 1991 coup. He had specialized in Russia, dealing with the aftermath of Alexander Litvinenko's murder in London. In 2009, he left MI6 and, with a former colleague, founded a private business intelligence company. It had been subcontracted by a US company, Fusion GPS, which had been asked to carry out research on Donald Trump's links to Moscow, first for a Republican candidate and then for the Democrats. Steele began to tap his sources and wrote a series of reports—never intended to be a single "dossier"—on what he found. It was raw intelligence, not written for public consumption, and not every element would be corroborated. But it seemed to suggest links between the Trump campaign and Russia that went far deeper than anyone understood. He became so alarmed by what he saw that he passed the material to contacts at the FBI. In the aftermath of the election, he was worried about the impact on Britain's security (including the possibility that any British sources in Moscow could be at risk if details were shared with the Americans) and so contacted officials in London, including Charles Farr. Farr had run operations against Moscow in the 1990s and risen to become the chair of the Joint Intelligence Committee. Details would also be passed to the police and MI5 but British officials would remain silent, fearful of being drawn into American politics. On the American side, the dossier and Steele's role would become the subject of bitter partisan debate after it was published on the eve of Donald Trump's inauguration with questions as to why a foreign national was carrying out research on American politicians. Former FBI director Robert Mueller would in 2019 report that he found no evidence of collusion by the Trump campaign but plenty of evidence of Russian interference.

RUSSIA INTERFERED IN the elections but the much harder question to answer is what difference it made. The social media operations were extensive—Russian Facebook posts are estimated to have reached 126 million users. But how many minds did they change? That is harder to measure. Much more of the division was homegrown. The hack-and-leak operations surrounding the DNC and the Clinton campaign may have been more effective than social media influence operations. While it is hard to judge the precise impact on voting, one thing Russian activity did do was attract enormous attention. One result of the hacking, dossiers, and cyber illegals was that suddenly awareness of Russian activity would go from near nothing to all-consuming and the focus of intense political infighting as to its true nature. In this sense, Russia would achieve one of its goals in dividing its adversary and keeping it off balance, playing on existing political splits. It would also create an image of a Russian intelligence operation that was actually more powerful and coordinated than the often messy reality. From being barely aware of Russian activity, people would start to see its hidden hand everywhere. Russian ghosts would start to haunt the American body politic.

AFTER THE LEAKED DNC emails first appeared, Russian officials told journalists in Moscow that there was nothing to the allegation that the Kremlin was behind it all. One commentator made some rather perceptive comments about Western politics in July 2016: "We see that in the West, including the United States, people no longer trust the elites. . . . In the UK, there has already been a protest vote—Brexit. Now Clinton could suffer from a protest vote. People will not vote for Trump, but against Clinton and the elites." The commentator was Andrey Bezrukov, formerly Donald Heathfield. Clinton was in a much worse position than the polls suggested, he argued, and she had failed to see that her links to Wall Street were damaging. The former illegal would be proved right where many US commentators got it wrong. He would later cowrite a profile of Donald Trump, saying Clinton and Obama symbolized "the end of the Cold War political culture." "Trump's historical function was to shatter the United States' old, inaccurate, politically correct picture of itself. . . . Only an outsider can accomplish this," he and his coauthors wrote. Bezrukov was a man who understood both sides in a way few others did. The old illegals had been busted but the new online illegals had embedded themselves in American society in a different way. It was a sign of Russia's persistence in its aims but also its flexibility in its methods of pursuing them. It had developed a multifaceted campaign that used hacking, social media, cutouts, and businessmen. As well as trained intelligence officers, there were other types of Russians inside the United States, one of whom would come to be seen as the "new Anna Chapman" even though she was actually something quite different.

27

The New Ways

IN THE EARLY hours of November 9, 2016, as news of the victory of Donald Trump was making its way around the world, Maria Butina was busy on Twitter. Butina was an attractive redheaded Russian student in her twenties, living in America. She was sending direct messages to Alexander Torshin in Moscow. Torshin was deputy head of the Russian Central Bank. Butina suggested they talk on the phone. But Torshin was reluctant: "All our phones are being listened to!" he said. Eventually she signaled it was getting late: "I'm going to sleep. It's 3 am here. I am ready for further orders." Two days later, she asked Torshin to find out how "our people" felt about her prediction of the person who would be appointed secretary of state. Next she sent a series of proposals about establishing a dialogue with US politicians through a conference.

A year and a half later, in July 2018, the FBI pounded on the door of her apartment in northwest Washington, DC. Butina, who had just finished a master's at American University, was handcuffed and led away. But what was Maria Butina? "If I'm a spy," she told journalist James Bamford, "I'm the worst spy you could imagine." After she was arrested, Moscow said she could not be a spy because she had always acted entirely openly. But the FBI alleged that she had come as a "covert Russian agent" directed by Torshin as part of an "influence operation." The intention, it was alleged, was to infiltrate US political organizations to advance Russia's interests. One counterintelligence official says they were aware of at least half a dozen cases like Butina. But she was the one the FBI had decided to prosecute and bring into the public domain—the intention being to raise awareness of Russian activity in the United States and how it evolved. While superficially there were echoes of the illegals and especially Anna Chapman, the reality was that this was something different.

Butina had grown up in Siberia before moving to Moscow. She became involved with the Young Guard of United Russia, the youth wing of Vladimir Putin's party. She began traveling to the United States soon after she started to work as a special assistant to Alexander Torshin. He was then a senator for United Russia and had been building up contacts on the right of American politics beginning in 2009. The CIA had reportedly believed he had been doing so to advance Moscow's interests and in 2018 he would be one of a group of oligarchs and senior officials sanctioned by the United States. In 2012, he acted as an election observer in Tennessee, a useful chance to see how the system worked. Torshin invited US politicians back to Russia as he began to build his contacts.

In 2013, Butina met a conservative political activist. She was in her twenties and he was in his fifties but the two would end up in a relationship (he has denied any wrongdoing). By March 2015 she was emailing him from Russia with a plan called "the Diplomacy Project." This, the US government alleged, aimed to establish unofficial lines of communications with people influential in American politics for the benefit of the Russian government. The Republicans were likely to win in the 2016 election, it was thought, and normally took a tougher line on Russia. Her plan was to strengthen links with influential individuals before the election. Butina had a way in. Her Siberian upbringing meant hunting and guns were part of her life. In 2011, she had founded "Right to Bear Arms," a Russian version of America's National Rifle Association (NRA). She already had contacts in the NRA and had attended events where she had been introduced to Republican leaders as a "representative of informal diplomacy" in Russia.

In her communications there were references to a "funder" who was a well-connected Russian oligarch worth $1.2 billion. She sent her proposal to him and asked for a budget of $125,000 to participate in upcoming Republican Party events. In April 2015, she attended the NRA convention and reported back, as she did for future events. She invited members of the NRA back to Moscow, where they met senior officials, thanks to Torshin. Butina and her supporters have always argued that she was simply acting to try to improve relations between the two countries.

The FBI claimed her contact with Torshin involved "taskings" of the type agents receive. Butina told him she had identified a specific candidate she thought might win in 2016 and said she had managed a "short personal contact" with the candidate as well as some of his advisers. Torshin asked for more details. Butina sent a report along with an election forecast. He asked if he could send it on to the Ministry of Foreign Affairs. In June 2016 Butina received a visa to study and she arrived in the United States in August.

Her American conservative activist introduced her to contacts, explaining it would be easy to meet people if she could present herself as a potential line of communication into future Russian governments. Butina also discussed with another American plans to establish an "informal communication channel" by holding "friendship and dialogue" dinners with influential people. These would help "correct" the outlook toward Russia and avoid too "oppositional" a stance from the next administration. Butina said the Russian presidential administration had expressed approval "for building this communication channel." The FBI alleged this activity "could be used by the Russian Federation to penetrate the U.S. national decision-making apparatus to advance the agenda of the Russian Federation." Once the election was over, Butina told Torshin the next step would be for a handpicked Russian delegation to come to the February 2017 National Prayer Breakfast to develop a back channel. Butina's profile was growing. Reporters were asking who this unusual Russian was who moved in Republican gun-rights circles. Suspicion of Russia was on the rise amid talk of election interference. She and Torshin did not want publicity but still found some of it amusing.

"Are your admirers asking for your autographs yet? You have upstaged Anna Chapman," Torshin wrote to Butina in March 2017, according to the FBI. "She poses with toy pistols, while you are being published with real ones." In July 2018, Butina was arrested. The FBI was adamant she was working on behalf of the Russian state. In December 2018 she pleaded guilty to acting as an unregistered agent of a foreign government. The FBI claimed she was likely in contact with the FSB, but one note seems to suggest only that she had been approached by them and was unsure what to do. Evidence tying her directly to Russian intelligence was thin and far from conclusive. There was no sign of clandestine communications. The contacts with Torshin were over Twitter direct messages—hardly the stuff of steganography. Butina's lawyers said she simply shared her "naïve, youthful optimism for better Russian-American relations" with others. Her attendance at conferences, networking, and organizing of dinners was consistent, they said, with her own beliefs and done on her own initiative to improve relations. There was no accusation that she was an "operative" who stole secrets or covertly plotted against the United States, they added. She hardly hid her pro-Russian views, those who met her recalled, even alluding to her contacts with people she met on campus.

The FBI told a court she should be considered "on par with other covert Russian agents." But her communications are open to different interpretations. They certainly suggest efforts to keep a low profile. But do they—and the other evidence—show she was an actual spy? Perhaps not in the traditional sense. She was certainly not a trained SVR officer like the illegals. Nor was she even a "special agent" or "true name" SVR illegal in the manner of Anna Chapman, despite some superficial similarities. "It is not an element of any of the charged offenses that the defendant worked for a foreign intelligence operation," the US government said in court documents. But that, US counterintelligence officials say, is because Butina is an example of how Russian activity had continued to evolve. "Butina's work involved building a rolodex of information about powerful people who had, or were likely to get, access and influence over the next presidential administration," a sentencing memorandum in her case argued, adding, "Butina's reports back to the Russian official on the people she was meeting have all the hallmarks of spotting-and-assessing reports." In other words, even if she was not an actual spy working directly for the Russian government or even a fully conscious agent, she was doing the kind of work Heathfield and Murphy had been undertaking before her. The information she provided was of "substantial value" and the Russian intelligence services would be able to use it for years to come to identify and target those who might be susceptible to recruitment by trained intelligence officers, according to a former FBI official's declaration. He argued that her reports had "all the hallmarks of targeting packages used in spot-and-assess operations" focusing on people with political influence. In other words, Butina could be used by Russian intelligence even if she was not a spy herself.

IN THE PAST, Western spy-catchers saw the Russians work along lines that had barely changed over the decades with the same kind of tradecraft, such as illegals and regular spies under diplomatic cover. But in the last few years, they have begun to improvise much more. There appears to much freer rein to try things and see what works. Russian spying was becoming more opportunistic—more tactical. If the Russians see an opportunity, they grab it, is how one recently retired US counterintelligence official puts it.

Already during the Ghost Stories investigation, the FBI could see the evolution away from illegals who required years of investment and toward the Anna Chapman model—with lighter or minimal training. The time of the deep-cover family illegals had passed—they were expensive and 2010 had shown their limits. Special agent illegals like Chapman worked for the SVR but had less cover—they were openly Russian. But a further evolution was the increasing use of "co-optees." They were Russians who were not trained spies at all but who were co-opted in to help. In some cases, these people might not even have known they were working for Russian intelligence. The advantage was they provided more plausible deniability—if they were caught they (and Moscow) could say they were just a student. If you catch an illegal and can prove they are using a fake identity, it is over. The SVR uses this approach but so does the FSB and other parts of the state. The other hard reality is that those who are co-opted are "disposable"—they are not trained officers with secrets in their head who need to be retrieved (or swapped). If they end up in jail, that is embarrassing but less of a problem for the intelligence services.

The ability to run co-optees and special agents with less training is also aided by technology. In the past, communicating with Moscow securely required tradecraft training on invisible ink or steganography. But now you can use commercial encryption systems that everyone can download. A decade ago, using high-end, military-grade encryption was enough to get you noticed by security services. Now commercial encryption is so widespread, it is much easier for a nontrained co-optee to use it and not look suspicious, hiding amid the noise. This model involves less investment and less risk. It is also much better suited to a model in which your goal is influence operations rather than classic espionage.

Another change was the role of oligarchs. It is telling that both Maria Butina and the Internet Research Agency were, the United States alleges, linked to oligarchs and businessmen. This is another shift. When Putin came to power the oligarchs had been told to bend the knee. You were not allowed to survive as an oligarch unless you worked with the state (which is not quite the same as working for the state). A decade or so later, as the Kremlin turned to influence campaigns, the oligarchs were now ideal tools to work through abroad. They had two advantages. One was that there was a level of deniability to their actions when discovered. The Kremlin could always say that these were people acting on their own in a way they could not when an illegal or a GRU officer or a diplomat was caught interfering in politics. Second, the oligarchs had deep reach into the West. These were the men (and they were all men) who owned football clubs and sponsored art shows, who had politicians over to stay on their luxury yachts or at their Italian and French villas. Increasingly oligarchs have been used for both intelligence collection and influence operations, counterintelligence officials say. This includes offering business deals to Westerners as a means of securing influence over them (sometimes even just the possibility of a deal is enough to lead to a change in behavior). This was part of a wider shift toward using "cutouts"—the United States alleges this was how they started to spread some of the hacked material in 2016, not just through WikiLeaks but also fronts like online hackers such as Guccifer 2.0. In the past, Russian intelligence agencies would have wanted control of such activities to be kept "in-house" for reasons of secrecy.

But now this was changing. In practice, the Kremlin only needed to set an overall direction and could then allow oligarchs and spy agencies to all compete to please their master with relatively little oversight.

The new ways of Russian spying were more opportunistic, more fast and loose. Rather than sending agents out like the illegals or Anna Chapman, if the state sees a Russian in a position of influence, then it can approach them to recruit or co-opt them. It could be a student or a businessman. They may work wittingly or unwittingly, perhaps not knowing who they are ultimately working for. It moves away from some of the traditional concepts of sleeper cells who embed themselves and try to slowly work their way into circles of influence. Sometimes they are not even Russian—Directorate S has always had Department 10, which looks at people like foreign students or businessmen or women in Russia to see if they can be recruited to work for them. The results can be impressive. However you view Maria Butina, she had managed to build up a far larger repertoire of American political contacts than the deep-cover illegals had managed and in a far shorter time. The Kremlin could now work through a much wider range of outlets than just the spy agencies but use the deeper contacts with the West in the Cold War era—through businesses and the like—to seek out those with influence.

THESE NEW, CHEAP, fast, flexible agents were good and bad for the FBI. The good side was that more opportunistic and more tactical almost certainly also means more sloppy and more likely to make mistakes. The bad news was that it challenged the way counterintelligence had traditionally been done. The FBI excelled at developing sources and recruiting agents like Poteyev, bugging embassies and homes and following diplomats around Washington, DC, New York, and San Francisco. But that was based on a model of traditional intelligence officers who communicated with Moscow Center and could be identified and tailed. Now ordinary people were co-optees—perhaps working through cutouts, loosely connected to the state, and with not everyone considering themselves to be a spy. That was harder to find and harder to prosecute. One former senior FBI intelligence officer describes it like the parallel challenge in dealing with the transition between the old model of counterterrorism in which you traced cells connected back to Al Qaeda leadership and the new model in which you have ISIS cells that are self-starting. There are many, many Russians in the West and only a tiny percentage of them will be co-opted. But working out who they are is not easy.

British officials say they also see this new and evolving Russian tradecraft but see it alongside everything that came before rather than replacing it. They still see spies under traditional or diplomatic cover, or nondiplomatic (working for, say, Aeroflot), and also illegals who have never disappeared. They still see old-fashioned dead-letter boxes—although now it may be more likely a sticker than a chalk mark to signal a site needs to be cleared. The Russians will not get rid of anything that worked for them, reckons one. There are certainly changes even in the old ways of spying, though, particularly thanks to technology—you may not need to give someone a precise description to find a drop site if their phone can ping when they are near it. But now Russia has more options—new ways of spying and using Russians and technology around the world.

THE US ELECTION of 2016 and the growing exposure of its new agents and methods did not end Russian activity. Immediately after the votes had been counted, the cyber actors started sending malicious links to US government employees and individuals associated with US think tanks and NGOs in national security, defense, and foreign policy fields.

In St. Petersburg the pace actually increased. The highest peak of Internet Research Agency ad volume on Facebook was in April 2017 but it continued into 2018. It sought to divide by ramping up the rhetoric on divisive issues like race relations, immigration, gun control, LGBT rights, and the NFL national anthem debate. It would take advantage of events like shootings in Parkland, Charleston, or Las Vegas or of African Americans by the police to increase the temperature. It also began to react to the inquiry into Russian interference itself. Specialists were told to react to one article about Robert Mueller by saying, "Mueller is a puppet of the establishment. List scandals that took place when Mueller headed the FBI."

The new cyber illegals were also still out there, such as supposed New Yorker Bertha Malone on Facebook, demanding to "Stop All Invaders," her content reaching nearly 1.4 million individuals in just a few days. Russia had found a system that it believed worked and it was not going to stop. But the newfound aggression was not confined to what it was doing online. For Moscow, it was also expressed in its desire to settle old scores, scores that went back to the destruction of its American illegal network. Murder was on its mind when it came to chasing those ghosts of the past.

28

Revenge

IT WAS MIDAFTERNOON on March 4, 2018, when emergency services received a phone call from members of the public in Salisbury. There was a strange couple—an old man and a young woman—ill on a bench in the cathedral city. "The lady was sort of passed out and leaning on the guy and he was doing some strange hand movements," a passerby recalled. "They looked totally out of it." Some people thought they were addicts or drunks and simply walked past. But one or two noticed their clothes did not quite fit that description. An ambulance whisked them away to a hospital. Heavily sedated, their lives hung in the balance.

A CALL CAME in to the duty officer at MI6 headquarters at Vauxhall Cross that evening. The duty officer is the overnight port of call for any emergencies, staying in a small apartment with three TVs and four phones. The officers are normally "old warhorses" of the service who have been around the block and are reaching the end of their careers but who know how to handle the unexpected. They receive a briefing document on the twenty or thirty things that might make one of the phones ring and which desk officer to contact when it does—perhaps an agent who had missed a meeting or, at the most dramatic, someone who was compromised and needed to get out fast. Most nights pass without incident. But that evening's call was not covered by the list. Local police in Salisbury had identified the man who had been taken into the hospital. They had done a Web search on his name. And that had set off alarm bells. He was a Russian spy who had been living among them.

SALISBURY IS A picturesque, quintessentially English town, an incongruous place to find a former Russian paratrooper and spy. But it was where Sergei Skripal had made his new life. Neighbors were oblivious to the fact that the man barbecuing sausages to the point that they were largely inedible in the garden of his modest home had fought and killed in Afghanistan and exchanged secret messages with British spies. But it was also hard for those who met him to miss the fact that he was a Russian. Any hopes of a quiet, peaceful retirement would prove illusory. First, there had been domestic tragedy. In 2011, Skripal's wife was diagnosed with cancer. After she died, he would still talk to her as if she were there. His son, Sasha, still living in Russia, had suffered from the stigma of treachery since his father's arrest. He turned to drink and developed severe liver problems. In July 2017, he died at age forty-three. His body was brought to the United Kingdom and Skripal would visit his grave after his weekly shopping trip, carrying a sense of guilt. Skripal's brother died in 2016 in a car crash. It was as if death were stalking him.

THE LIFE OF spies after they defect or are swapped is rarely easy (much has been made of Kim Philby's time in Moscow listening to cricket matches on the radio). They are a "declining asset" in terms of their value over time, one person who looked after them explains. But those who knew him say Skripal was resilient, enjoying TV documentaries and visiting local museums with a military flavor like the tank museum in Bovington and the Army Flying

Museum in the delightfully named village of Middle Wallop. He did miss Russia. He would buy Russian dumplings in bulk and freeze them, eating them while playing World War II tank games on his computer. He gave the odd lecture about the GRU to military academies and European spy services, as former agents often did. But any knowledge would have been dated, given he had left the GRU two decades earlier.

By spring 2018, Sergei Skripal had lived through much—from Kaliningrad through Afghanistan to Vienna and Salisbury. He was looking forward to the visit of his daughter Yulia, who arrived on Saturday, March 3. On Sunday, they left Zizzi restaurant just after three thirty. It was a popular place with the family. A picture of Yulia and Sergei, her holding up a glass of wine and him, looking happy, holding up a half-drunk pint of lager, would become one of the iconic images of the pair in the coming days. What few realized was that in the mirror behind them you can see the reflection of the person taking the picture, the departed Sasha, the photo taken on a visit before he died. That Sunday afternoon after lunch father and daughter began to feel ill and then staggered to a park bench. Next they were in the hospital, hovering between life and death. But unlike Alexander Litvinenko, they would eventually pull through thanks to good fortune and remarkable medical care.

Inside the British Secret Service's headquarters, the realization that the Russians had targeted one of their agents on home soil sent shock waves through the building. It threatened the very core of their business. Protecting agents is vital. If those who chose to betray secrets to MI6 could not be protected in Britain itself, then others contemplating such an act might well think twice. This, of course, was precisely the Kremlin's intention.

Why was Skripal targeted? Various motives have been suggested, including that he might have been carrying out operational activity—perhaps in Ukraine or by identifying former colleagues to be approached. But it was simpler than that. He was a traitor, in the eyes of the GRU and of Putin. He was a military man who had gone against his oath for money. But it was not just about punishing him. It was also about sending a message to others—not so much those who had already walked down his path and were living abroad, but rather to those within Russia considering treachery. If you were an SVR or GRU officer or any other Russian official thinking of responding to that British or American pitch you had encountered, did you really want to take that risk? Even if you seemed to have got away with it and cashed out a few million dollars and had a home in Florida or Salisbury, your former colleagues would still come looking for you. Even after close to a decade, when you thought you were living a quiet life, they would still hunt you down and try to kill you and not care about hurting those you loved as they did it. This was the message. It was a message that made perfect sense for a regime that had built its foundations on spy fever. And, as with Litvinenko, the main piece of evidence pointing to the Russian state had been the use of the most unusual of weapons.

A CALL HAD come into Porton Down, home to Britain's secretive biological and chemical research establishment, early Monday morning. A rapid-response team was sent out to the scene within hours. The teams work 24/7 attending scenes of terrorist attacks to look for traces of explosives and chemical or biological weapons. For those responding to Salisbury, the most jarring element was that they were carrying out this search almost literally on their doorstep, in the same community where many investigators lived. Soon after the attack, a visit to the usually closed and secretive site revealed a strange landscape with odd-shaped buildings almost like Hobbit homes where explosives are tested, with red warning flags littered around. The site is purely defensive in its work today, officials stressed to me. Inside one 1930s building is a tightly sealed chamber where a robot marches on the spot and live "agent" is pumped in to see how long it takes to burn through the suit it is wearing. Such is the danger that when the chamber is washed out afterward, the tanks that hold the residue have to be carefully destroyed. After the Salisbury incident, samples were rushed back to the labs on-site. They identified a highly unusual poison—A234, a military-grade nerve agent from the Novichok family. It was developed by the Soviet Union specifically to get around NATO's defenses.

The revelation from Porton Down added to the sense of shock. A policeman wearing gloves had opened Skripal's front door, but the Novichok still made it through and poisoned him. It had been smeared messily on the door handle. The perfume bottle it had been carried in was dumped callously in a charity donation box, killing a local woman, Dawn Sturgess, who picked it up and sprayed herself with it three months later.

The chief of MI6, Alex Younger, made one of his rare public appearances at St. Andrew's University in Scotland six months after the poisoning. I asked him whether he thought MI6 had failed to do enough to protect its former agent. Younger began with the boilerplate response that it was impossible to confirm or deny who might have been one of MI6's agents or what arrangements are put in place to protect them. But then his face darkened. "Mr. Skripal came to the UK in an American-brokered exchange, having been pardoned by the president of Russia," he said in response to my question. "And to the extent that we assumed that had any meaning, that is not an assumption we will make again."

The poisoning was a failure, several senior officials who served in British intelligence concede. A risk assessment was carried out when Skripal was swapped out. The view was that he had been pardoned and should be safe and so would not need to be given a new identity. As a "settled defector" he had his own say on what kind of security he wanted to live under. But veterans say that the pardon should never have been seen as counting for much since the Russians would have felt they were forced into it. "People with my background don't believe in things like that," says one former spy. Putin, they say, will have bitterly hated being forced into a swap. And the Putin of 2018 was a different Putin from the man he was a few years earlier. There may have once been an unwritten rule that spy services do not go after swapped defectors. But anyone who had been watching Russia should have realized that one thing was now true—Putin no longer played by the old rules.

Skripal's risk assessment was never updated to take into account that the Russia of 2018 was very different from the Russia of 2010. British national security officials express surprise at the lack of protection around the spy. There were not even security cameras on his house to watch anyone who came to visit. MI5 and the Home Office did carry out a "refresh" on the security of defectors after the fact, something officials acknowledge was overdue. MI5—now three times larger than it was in the 1990s—has shifted more resources toward Russia. But it still sees counterterrorism as its core mission, and it has been as busy as ever stopping terrorist plots. Russia was long described as an annoying distraction from the priority of saving lives. But as evidence of Russia's willingness to interfere politically and kill people on British soil grew, that argument seemed harder to sustain. Suddenly the spying game looked a lot more serious and deadly.

Detective work would identify two GRU officers who came to Britain as the main suspects. The ease of their identification made the point all too clearly about how spying and maintaining your cover had become much harder in the digital age. A combination of passport data and CCTV images allowed a private investigative group, Bellingcat, not only to name the two officers but even trace dozens of other GRU operatives and their travels. The bizarre appearance on Russian TV of the pair—in which they claimed that they had come to Salisbury as tourists to admire the spire of the cathedral—was met with widespread derision in Britain. The pair, it is fair to say, did not look like tourists who would have come all the way to the United Kingdom for a weekend to see a spire, however impressive. As with Litvinenko's alleged assassins, it was tempting to see Russia's spies as bungling and inept. Until you remember that in both cases people died. Salisbury was characteristic of their new approach—highly aggressive, even reckless, in intent but also often sloppy in execution, perhaps reflecting a lack of care about getting caught. After all, even if you are caught, then you can deny everything, as Russia has done, and use your information warfare tools to muddy the waters while also knowing the effect will still be to intimidate.

RUSSIA HAS LONG had a program to go after those seen as traitors, says one senior official. In the Cold War, it drew up lists of those singled out for assassination. Embassy officers were given the job of establishing their whereabouts and movements, down to what pub they liked to frequent. Department 8 of Directorate S had the task of carrying out any actual killings, though. Over the past ten to fifteen years, officials say there have been at least a dozen cases of individuals being targeted—mostly within Russia or around its borders. Russians abroad have died in mysterious circumstances in a multitude of cases. In some of these cases, there may be nothing suspicious. In others it may be linked to corruption and organized crime, and still others to treachery. But often the investigations have been so cursory that the real answer may never be known, which leaves you wondering if governments may sometimes prefer not to find out and have to deal with the consequences. Russian exiles frequently complain of surveillance by operatives in the United Kingdom or express fears that potential agents may be inserted into their circle of acquaintances in order to report back.

Why was Skripal targeted in March 2018? Perhaps because Britain seemed weak. Its response to the killing of Litvinenko had been pathetic. And it was on the back foot because of Brexit. That was likely how Moscow saw things. The chances of a tough response, they thought, were low. Those with a close insight into the investigation say the assassination attempt had been a long time in the planning—at least a year. The British government said that the GRU had been hacking into Yulia's email from as early as 2013, likely for reconnaissance, something that was only discovered when forensics were done on her machine after the attack. The final timing may well have been opportunistic based around tracking Yulia's travel or when the assassins were available, given that they look to have been busy carrying out operational tasks in Ukraine just before.

Putin himself never admitted Salisbury was a Russian operation. But when he was asked about events just over a year later, he reiterated his absolutist view on treachery. "Treason is the gravest crime possible and traitors must be punished. I am not saying that the Salisbury incident is the way to do it. Not at all. But traitors must be punished. . . . It is the most despicable crime that one can imagine." How high up was the operation approved? At a senior level, British officials say, although they cannot be sure Putin gave his explicit go-ahead. One possibility is that there was a standing order to go after traitors and the opportunity simply came up and was taken. Elements of the Russian state—including the GRU—have wide latitude to act within broader orders from the Kremlin and will, on occasion, compete to show how aggressive they are.

There were reports that former top SVR officials had got in

touch with European counterparts shortly after the poisoning to explain they were not responsible. It was a sign of the fallout and infighting—many suspected that information that emerged about the GRU's role came from the FSB, keen to portray the GRU as a bunch of useless thugs who had been sloppy so they could gain the upper hand in foreign operations.

London had learned something from its experience with Litvinenko and the reaction was tougher this time. Every known Russian intelligence officer operating under diplomatic cover in the United Kingdom (apart from the declared liaison officer for each service) was expelled—twenty-three in all. They came from all three Russian intelligence services, but British officials say the SVR officers were the most active talent spotters and recruiters of agents and the ones who caused most concern when it came to political influence, cultivating contacts, and looking for sources. The decision was taken to kick out the spies fast—part of the reason being to give them less time to hand over agents they were running and make it easier to detect either those agents or any interim case officers who were going to handle them (they might be working as illegals under nondiplomatic cover). At the British National Security Council meeting discussing the move, the issue of retaliation was raised. The chief of MI6 almost laughed when it was asked whether he had twenty-three officers the Russians could expel in Moscow. The real number was nothing like that, which meant a fair few genuine diplomats would have to pack their bags. Russian spies are nervous about being kicked out of a cushy posting like London, since it leads to being blocked from anywhere in the West. British officials believe they did real damage to Russian intelligence operations in the country but are under no illusions that there are not more spies still at large. And for all the talk of tough action, there was still no significant move against all the corrupt Russian money swilling around London, no denial of access to financial markets for Moscow or restrictions on the purchase of property through shell companies. And it did not take very long for the voices to resurface arguing that it was time to forget all the spy stuff and get back to business. Old habits die hard.

IN WASHINGTON THERE was surprise at Moscow's audacity. "It is as if we were to take the time to go after the poor [illegal] families and knock them off. It is crazy," says Leon Panetta, the former CIA chief who oversaw the Vienna swap. "It's a reflection that Putin never forgot what happened here and wanted to make a point." The United States expelled sixty Russian diplomats, twelve of them at the UN. A senior US official said at the time there were "well over" a hundred Russian spies under diplomatic cover, probably around 150. The Russian consulate in Seattle was shuttered (it was reported that Microsoft had been a target as Russian spies looked for employees and coders who could be recruited). It also emerged that some of the Russian facilities dotted around the country that were closed down at the end of 2016 in retaliation for election interference had been used for an operation that had managed to monitor the secure communications of the FBI's surveillance teams after the end of the Ghost Stories investigation.

The expulsions and closures damaged Russian operations in the West, but the reality is that throwing out intelligence officers under diplomatic cover makes far less difference than it did in the past. In the Cold War, spies under diplomatic cover and illegals were the primary way the Russians could recruit and run agents and steal secrets. Now there are many more ways. There is cyber espionage, allowing the SVR, GRU, and FSB to steal secrets remotely from behind a computer screen in Moscow. The online world has also transformed human intelligence. In the past, it would take legwork by agents or illegals to research a potential target. Now people post on Facebook or tweet. Spies can then use covert access to databases to further build their picture. And the Russians have new options like business cover or people co-opted through cutouts and oligarchs to work on the ground. Russia has also increasingly turned to spies on short-term visas, who are less likely to be the subject of intense surveillance than counterparts operating out of embassies. These "TDY" travelers (military slang for someone on temporary duty) are used increasingly to clear dead drops and collect intelligence. Both British and American security officials say they regularly see spies heading out to the countryside with rucksacks to map out communications infrastructure—satellite links, ground stations, and internet exchange points. They cannot be sure whether it is for espionage—where to put a probe to siphon off data—or for illegals to carry out sabotage in time of war. These spies were also used to carry out reconnaissance on defectors in the United States in the wake of the 2010 arrests, alongside embassy-based counterparts.

AFTER THE SKRIPAL case, the FBI reviewed its own protection of defectors. Sitting in his comfortable suburban home, a former KGB officer in the United States explains he had always assumed that the long arm of the Kremlin would not reach into America. But would they really not risk the consequences? Not everyone is sure anymore. The Skripal case was just one sign that this assumption might be outdated. After Salisbury, urgent inquiries were made into how easy it might be to find defectors and former agents in the United States. Hundreds of individuals and their families are looked after by the National Resettlement Operations Center, within the CIA. Its staff have always taken extensive precautions, using Moscow-level tradecraft in Washington to make sure they are not followed when they leave their anonymous office to meet defectors for a catch-up. But it is getting harder to protect them now, CIA veterans say, partly because their families want to be on social media. Defectors are normally advised to keep a low profile but even if they take that advice, disappearing is getting much harder to do thanks to their digital trails. Some also share that risky characteristic of spies who lived on the edge—they are sure they will never get caught. The Russian program to track down defectors was boosted around 2014—exactly the moment that its other operations, whether cyber hacking or political interference, were also being unleashed. It was all part of the conflict that the Russians felt they were already engaged in but which the West had not fully appreciated.

In July 2016, the man who was Russia's number one target was reported dead. Russia's Interfax news agency reported that Alexander Poteyev had passed away, basing its report on "informed sources." "Good riddance," was the response from Andrey Bezrukov. "It is known that people do not live long in such a state," Bezrukov went on to say. "A person who has committed treason is under enormous stress. He lives with it every day. He is lonely. . . . Such people often take to drink. They commit suicide. They lose the purpose of life."

But Poteyev was not dead. One theory is that it was an American ruse to keep assassins off his trail. Or perhaps a lie by the Russians to show their public that a traitor had met with what he deserved. Another theory is that the Russians were putting out the story in the hope he would break cover to contact relatives, providing potential intelligence on his whereabouts. Poteyev, though, was not hard to find and already had a habit of calling people back home despite repeated warnings from his CIA minders of the risks, according to one official. At the start of 2018, a senior US intelligence official warned against asking too many questions about the former SVR man's whereabouts. When, two months later, Skripal was nearly killed, it became obvious why the warning was so stern.

• • •

THE SKRIPAL ATTACK might have been an indication of how far the Russians were willing to go, but even before it there had been warning signs that Poteyev had not been forgotten. More than three years earlier, the FBI picked up a Russian operative entering the United States. He was traveling on a visa rather than operating out of the embassy, but the United States was aware of his identity and able to follow him. Surveillance teams realized he was looking for Poteyev. The operative traveled to the former spy's house. It was part of an established but increasingly aggressive Russian program to establish the location of defectors and there was a debate about how worried to be. Some voiced doubts as to whether Putin would actually do anything on American soil. At the time, the judgment was that the mission was reconnaissance rather than a planned hit and the Russian was watched rather than stopped to gather intelligence about his activities, even as he approached the home. But it may have been a closer call than has been previously understood.

After Skripal's poisoning, the British government revealed it had received highly sensitive intelligence that pointed clearly to Russian involvement in Salisbury. They had learned that from the 2000s Russians had begun training special teams of assassins to deliver chemical weapons. This program later looked specifically at smearing nerve agents on door handles. This intelligence, it can be revealed, came from the United States. One national security official says that it came because the Russians had been spotted trying to do something similar to Salisbury inside the United States. Another official describes Russia going so far as carrying out what they call a "dress rehearsal" for the Salisbury poisoning in the United States. A nerve agent was not used but operatives wanted to see if they could approach a door to be able to smear something on and carry out such an attack. The target may well have been Alexander Poteyev—although it was possible it was another individual in the US. The FBI and other US officials decline to comment on the details, citing security sensitivities. Whatever the case, it seems as if warning signs about Russian targeting were not well shared or understood among allies. One US counterintelligence official concedes that in the aftermath of Salisbury, they have been ratcheting up the assessments of Russia carrying out such an attack in the United States and preparing for what they describe as the day "when—not if" it comes.

IN THE WAKE of Salisbury, Poteyev, as with others who had been swapped, vanished. And so the man who began the Ghost Stories investigation and helped find Moscow's spies in the suburbs is now himself a Russian living somewhere in America and hovering between real life and an already-reported death. Another Ghost.

BUT ARE THERE illegals still out there?

THE GHOST STORIES investigation did real damage to Russian spying. It wasted decades of investment and led to the SVR expending energy on a program under FBI control. It helped the United States understand Russian tradecraft and helped allies around the world pursue leads in their countries. It damaged the prestige of the SVR. More important was what the investigation made sure did not happen. If Donald Heathfield or Cynthia Murphy had managed to recruit a next generation who could pass background checks and get jobs within the American intelligence community or government, then the damage could have been enormous—perhaps comparable to the pain inflicted by the Cambridge spies recruited by illegals in the 1930s. "We believe we eviscerated that [program] back in 2010," Bill Evanina, director of the National Counterintelligence and Security Center, said of the illegals in April 2018.

US spy-catchers working in the wake of the 2010 arrests say they saw a shift away from illegals. The verdict from Putin, they believe, was that the old illegals had not justified their cost and the program had embarrassed him. It is certainly getting harder for them. A check on a database can spot a dead double. There are many new alternatives now—cyber illegals, Special Agent illegals, and others—to carry out their missions. But that does not mean the old illegals have vanished.

Putin was asked on TV in 2017 if he had a message for illegals serving abroad. He first made a reference to his own work supporting illegals when he was a KGB officer back in Dresden. Then he went on to address directly those living under deep cover: "I know what kind of people these are. These are special people, people of special qualities, of special conviction, with a special character. To give up their life, their nearest and dearest and leave the country for many years, and to dedicate one's life to the Motherland, not everyone is capable of doing that." Putin's performance was a deliberate attempt to tap into the mythology surrounding the illegals and to associate himself with it as an election approached. His comments revealed how illegals were still rooted deep within the psyche of Russian intelligence and popular culture. Do I think the Russians need them? one US counterintelligence official asks rhetorically. No. But will they still use them? Yes, because it is part of their culture, they argue. "This is what the Russians do—they will always do it," says Alan Kohler, who supervised the Ghost Stories investigation. "It may take different forms and use different technologies, but this is always what we are going to face."

The carefully managed strategy to make heroes of Russia's spies continued. News broadcasts as well as dramas are filled with positive stories about the work of the FSB (a prize was created in 2006 for the most positive depictions of the work of a state security officer in culture). As a result, the percentage of Russians approving of the work of their own intelligence services went from 35 percent in 2001 to 66 percent in 2018. After the success of a Western TV drama on the Chernobyl nuclear disaster that focused on the lies pushed by the government, a state-backed Russian TV channel announced plans for its own version in which the hero was a Russian counterintelligence officer tracking down a CIA agent operating around the plant. Andrey Bezrukov and Elena Vavilov began to appear on talk shows to discuss their lives. In one case they appeared alongside their parents, friends, a Russian actor who had starred in the drama *The Americans,* and an actress who had starred in the drama *Seventeen Moments of Spring* and who expressed her admiration for the modern-day illegals. The audience lapped it up. The only time the couple appeared uncomfortable was when they were asked what it felt like to have been watched by the FBI the whole time and basically achieve next to nothing in their years abroad.

In the West, Russian espionage and influence has also been front-page news and center stage on-screen, not least thanks to election interference and assassination attempts. And there are risks to that. Because just like those Western spies trained to watch out for Russian operatives on their heels as they walked through Moscow, the danger is that everyone becomes hyperaware and begins to see "ghosts"—the hand of Russia's spies—everywhere even when they are not actually present or as powerful as portrayed. But the old hands inside the intelligence agencies who have spent their careers watching their adversary know this attention does not always last. The media and the public's focus fades. New threats emerge. But even when that happens, they understand that, as the Ghost Stories investigation showed, Russia is patient and persistent in its espionage. "The same thing has always been going on and will continue to go on whether or not people are paying attention to it," says Maria Ricci of the FBI. Whatever the diplomatic niceties are, whatever resets come and go and whether or not public attention diminishes, the Russian drive to spy and to use intelligence to influence will remain. But Ghost Stories also showed that the techniques by which this is carried out do change. The old ways of building cover and legends are harder thanks to biometrics and databases and there are other ways now to achieve their goals. But finding a well-trained illegal remains an enormous challenge, a Western spy-catcher explains, and they offer something that a cyber spy cannot. That means despite the blow the Ghost Stories investigation dealt, there is no reason to think Moscow Center will give up on the illegals that they have run for a century. At least not while Vladimir Putin is in power.

On the evening of June 28, 2017, Putin went to Yasenevo for a special gala to celebrate ninety-five years of illegals. Standing at a podium in the main auditorium, he paid tribute to Directorate S as a "legendary unit" and gave a roll call of its heroes, adding that after he had finished speaking there would be a closed ceremony in which he would be conferring one of the nation's highest honors on an illegal who had "displayed valor and courage while fulfilling special missions in life-threatening circumstances." It was no accident, he said, that the motto of the illegals was "without the right to glory but for the glory of the nation." Other states might have spies, he said, but Russia had the illegals. They were needed more than ever, and he wanted to thank everyone who had preserved their traditions in what he said were "difficult years"—most likely a reference to the bust of 2010. Putin ended with a call to arms for those in front of him and for "the agents who are now serving abroad." "I wish you good health, good luck, and new victories for the greater good of Russia."

. . .

MAKE NO MISTAKE, somewhere—living in suburbia, picking up their children from parties, smiling at the neighbors as they water the hydrangeas—illegals are still out there. "We don't consider ourselves heroes," Elena Vavilova said looking back, nearly a decade after she had been swapped in Vienna. "We just honestly did our duty." Her husband also plays down his role. "I am an average undercover agent. Hopefully not the worst, definitely not the very best. You have never heard about the best ones. And never will."

ACKNOWLEDGMENTS

THIS BOOK IS the result of two decades of reporting on intelligence agencies in the West and Russia. Many individuals have shared their stories and thoughts, but often anonymously. This is my chance to express my gratitude to them. I would also like to thank my agent, Georgina Capel, and her team for their support and encouragement. My two editors, David Highfill in New York and Arabella Pike in London, have expertly guided me in the writing of this book and the teams at HarperCollins have been of enormous assistance. My greatest debt is to my family for their patience and understanding.

NOTES

INTRODUCTION

7 "The Western world can't bring itself": Alexander Kouzminov, *Biological Espionage* (London: Greenhill Books, 2005), p. 109.

1: THREE DAYS IN AUGUST

11 They could end up dead: The account of events surrounding the coup is based on conversations with a number of individuals serving in Moscow at the time.

12 The bulk of its half-million personnel: J. Michael Waller, *Secret Empire: The KGB in Russia Today* (Boulder, CO: Westview Press, 1994), p. 10.

12 But there was something tentative about the move: British ambassador Rodric Braithwaite's telegram to FCO, "Moscow, August 19: The First Day of the Coup," of August 20, 1991, https://nsarchive2.gwu.edu/NSAEBB/NSAEBB357/index.htm.

13 At the moment the coup began, the man in charge of the First Chief Directorate: Leonid Shebarshin's own account is published in *The Intelligence,* a collection of memoirs of spy chiefs edited by Todor Boyadjiev and published in Bulgaria by Libra Scorp in 2006.

15 "My experience tells me that this practise quite justified itself": Leonid Shebarshin, "The Illegals," *Inside Story,* BBC TV, June 10, 1992.

15 When the 1991 coup began, a legendary figure made a surprise reappearance: Kouzminov, *Biological Espionage,* p. 130.

17 destroying some of the documentation so it could not fall into the wrong hands: This is according to Vasili Mitrokhin, as recounted in the obituary of Yuri Drozdov, *Times,* July 26, 2017.

18 There was at least one undercover CIA officer among the crowd: Milton Bearden and James Risen, *The Main Enemy* (New York: Century, 2003), p. 506.

19 "Moscow is silent": Chris Bowlby, "Vladimir Putin's Formative German Years," BBC News, http://www.bbc.co.uk/news/magazine-32066222.

19 Shebarshin opened his safe and pulled out incriminating papers: Milton Bearden, "Requiem for a Russian Spy," *Foreign Policy*, June 18, 2012.

20 The "insular subculture didn't want to let go of the Cold War": Bearden and Risen, *The Main Enemy*, p. 454.

20 "Sorry, trip's off, young man": Sulick's account was published by the CIA in its *Studies in Intelligence* in 2007. https://www.cia.gov/library/center-for-the-study-of-intelligence/csi-publications/csi-studies/studies/vol50no2/html_files/CIA_Lithuania_1.htm.

22 "How naïve to believe the fall": Victor Cherkashin, *Spy Handler* (New York: Basic Books, 2005).

22 "The Soviet Union is no more but eternal Russia remains": Shebarshin's account in Boyadjiev, *The Intelligence*.

22 Dzerzhinsky's picture was still up on the walls: Waller, *Secret Empire*, p. 33.

23 Shebarshin, decades later and in ill health, shot himself: https://www.reuters.com/article/us-russia-spy-suicide/former-soviet-kgb-spy-chief-commits-suicide-idUSBRE82ToWZ20120330.

24 "For me, my country": Details of Heathfield and Foley and quotes by them come from emails from Foley and their interviews with Russian TV and newspapers. In this case https://tvzvezda.ru/news/qhistory/content/201931106-ftYfe.html.

2: THE BIRTH OF AN ILLEGAL

25 never left the family: Details of the real Heathfield and Foley families from https://www.thestar.com/news/gta/2010/06/29/stolen_identity_shakes_brampton_family.html and Michael Friscolanti, "The Russian Spies Who Raised Us," *Maclean's*, August 10, 2017.

26 "It was considered a big success": Kouzminov, *Biological Espionage*, p. 114.

26 Roughly one in ten attempts would: https://espionagehistoryarchive.com/2016/06/03/kgb-directorate-s-training-an-illegal/.

27 "For me to forget this is to be left with nothing": Bezrukov interview, "Why Spies Are Like Scientists," *Russian Reporter*, October 11, 2012, https://expert.ru/russian_reporter/2012/40/pochemu-shpionyi-pohozhi-na-uchenyih/.

28 Rudenko's career: Christopher Andrew and Vasili Mitrokhin, *The Mitrokhin Archive: The KGB in Europe and the West* (Penguin, London 2000), p. 253.

29 "You would not have to waste your time chasing after girls": John Barron, *KGB Today* (London: Hodder & Stoughton, 1984), p. 307.

29 Yelena Borisnovna and Dimitry Olshevsky: "Deported Spies Return to

Russia: Ex-Soviet Spy Chief Tells How They Would Be Handpicked," *Toronto Star,* June 12, 1996.

29 "Andrey offered me something out of the ordinary": Russian NTV interview, April 2019, "Новые русские сенсации": "Их сдал предатель."

30 "I believe we were selected": Email from Elena Vavilova, 2019.

32 evade a polygraph lie-detector test: Elena Vavilova interview on Tomsk TV, posted December 2018, first broadcast July 2017.

32 Each illegal required a staggering investment: Kouzminov, *Biological Espionage.*

32 "You do not train illegals . . . in the classes": Interview with Vladimir Semichastny for PBS Red Files, http://www.pbs.org/redfiles/kgb/.

32 Moscow had developed its specialty in illegals after the 1917 revolution: Nigel West, *The Illegals* (London: Hodder, 1993), provides an excellent overview of illegals.

33 atomic spies: Directorate S is believed to have been named after its founder Pavel Sudoplatov, who was given the mission during World War II of carrying out special operations and obtaining atomic secrets.

34 One of his first jobs in East Germany in the 1980s: Masha Gessen, *The Man Without a Face* (New York: Riverhead, 2012), p. 64; Fiona Hill and Clifford Gaddy, *Mr. Putin* (Washington, DC: Brookings, 2005), p. 276; and https://www.bbc.co.uk/news/magazine-39862225.

34 The CIA and MI6 have never had quite the same capability: In the Cold War some US companies regularly passed the CIA details of young men they recruited to work overseas to see if any could be of use. C. D. Edbrook, "Principles of Deep Cover," *Studies in Intelligence* 5 (Summer 1961).

37 "We trained authentic Americans and Englishmen on Soviet territory": https://espionagehistoryarchive.com/2016/06/03/kgb-directorate-s-training-an-illegal/ and also https://espionagehistoryarchive.com/2015/05/15/inside-kgb-directorate-s-the-illegals/.

37 When one illegal returned to Moscow: Kouzminov, *Biological Espionage,* p. 62.

37 even given a truth drug: Kouzminov, *Biological Espionage,* p. 107. Illegals were normally also sent on a training assignment abroad. These can last from weeks to months. They will cross borders in order to develop their cover story—perhaps by attending a short course undercover at a Western university—and undertake a set of tasks that also may help build up cover.

38 swear an oath to the party: The Mitrokhin archive, located at K-16 506, Churchill College Archives, Cambridge University, details one oath, although Elena Vavilova says she swore a standard military oath.

38 Another head of the directorate compared: https://espionagehistoryarchive.com/2015/05/15/inside-kgb-directorate-s-the-illegals/#more-181/.

38 As the moment arrived, Bezrukov and Vavilova: The exact date of their

arrival is unknown to the Canadian immigration authorities—perhaps because they still do not know what identity they arrived with, according to documents from the Registrar of Citizenship in Canada sent to Alexander Vavilov, dated August 15, 2014. However, the couple in an interview talk of a twenty-three-year mission that would date their departure from Russia to 1987.

38 They had cleared out every pocket: Elena Vavilova recounts this process in her loosely fictionalized novel published in June 2019. She acknowledges in one interview to promote the book that details up to their departure for the West are based on truth.

3: STRANGERS IN A STRANGE LAND

41 The monument was at a picturesque site in Canada: Bezrukov and Vavilova TV interview, "Новые русские сенсации": "Русские разведчики-нелегалы в Америке," NTV, April 24, 2019, https://www.youtube.com/watch?v=mDU879mQjSA&feature=player_embedded.

42 Canada was the ideal launching pad for illegals into America: In 1994 it was found that even secondary checks on suspicious individuals at the border picked up people who were not supposed to enter the country, with particular failures when it came to detecting false travel documents. It was only in 2002 after the 9/11 attacks that Canada improved its border intelligence and screening. http://www.oag-bvg.gc.ca/internet/English/parl_oag_200304_05_e_12911.html.

42 "Canada is a lot like the US, only colder": Jack Barsky, *Deep Undercover* (Carol Stream, IL: Tyndale Momentum, 2018), p. 137.

43 "I had to get an education again": http://rusrep.ru/article/2012/10/08/pochemu/ and https://www.pravda.ru/society/1426603-vavilova/.

43 "Every undercover agents' family have to decide": Joint interview of Bezrukov and Vavilova, https://tvzvezda.ru/schedule/programs/content/201705051916-tpaz.htm/201903010843-rtct.htm.

43 "We carefully weighed this, of course, discussed a lot": https://www.pravda.ru/society/1427554-vavilova/.

44 She had visited Hungary four years: James Adams, *The New Spies* (London: Pimlico, 1995) and "The Spy Who Came in from the Supermarket," *Daily Mail*, April 29, 1992.

45 increase the focus on illegals: Waller, *Secret Empire*, p. 135, and Kouzminov, *Biological Espionage*.

45 "strengthen the illegal branch": "The Illegals: Inside Story," BBC TV, June 10, 1992.

45 Alexander Kouzminov, later explained: Kouzminov, *Biological Espionage*, p. 135.

47 "the most detailed and extensive pool": UK Intelligence and Security Committee special report on Mitrokhin.

47 the defection meant that Moscow could not be sure that the legend: Christopher Andrew, *The Secret World* (London: Penguin, 2018), p. 715.

47 the two hundred leads Mitrokhin produced: Intelligence and Security Committee report 2003–2004 citing government response to its Mitrokhin Inquiry Report.

48 John Scarlett became MI6's first-ever head of station: Scarlett comments in a Royal United Services Institute podcast in September 2019.

49 The CIA and MI6 were opening up intelligence stations in its neighborhood: Yevgeny Primakov's memoir, *Russian Crossroads* (New Haven, CT: Yale University Press, 2004), p. 100.

50 "Russia as the defeated side": Trubnikov comments from interview. US secretary of state James Baker had used the formula of "not one inch eastward" in a 1990 meeting: GWU's National Security Archive report, https://nsarchive.gwu.edu/briefing-book/russia-programs/2017-12-12/nato-expansion-what-gorbachev-heard-western-leaders-early.

50 running a sting operation: Primakov makes a similar claim in his memoir, *At the Crossroads,* p. 101. The intelligence agency was Germany's BND.

4: "KARLA"

51 At that moment a young Russian walked up to him: Details of PROLOGUE and Zhomov drawn from a number of people directly involved in the case; also Bearden and Risen, *The Main Enemy,* and Sandy Grimes and Jeanne Vertefeuille, *Circle of Treason* (Annapolis, MD: Naval Institute Press, 2013), p 118.

57 code-named "Max": David Wise, *The Seven-Million-Dollar Spy* (Audible, 2018). The Russian would also provide intelligence that helped catch another traitor, Jim Nicholson. See Bryan Denson, *The Spy's Son* (New York: Perseus, 2016).

57 Aldrich Ames was a second-rate spy: Accounts are from a number of those involved and Grimes and Vertefeuille, *Circle of Treason;* Bearden and Risen, *The Main Enemy;* and Michael Sulick, *American Spies* (Washington, DC: Georgetown University Press, 2013).

58 A CIA delegation headed to Yasenevo to deal: https://www.belfercenter.org/publication/us-and-russian-intelligence-cooperation-during-yeltsin-years.

5: UNDERCOVER

61 "It is like walking a tightrope": "Новые русские сенсации": "Русские разведчики-нелегалы в Америке," NTV, April 24, 2019.

61 "Diapers Direct": Michael Friscolanti, "The Russian Spies Who Raised Us," *Maclean's*, August 10, 2017.

62 special boarding schools for the children of illegals: Kouzminov, *Biological Espionage*, p. 88.

62 In one case an illegal spent seventeen years abroad: https://espionagehistoryarchive.com/2015/04/01/interview-with-a-soviet-spymaster/.

62 "I was really young, I have no idea": https://www.theguardian.com/world/2016/may/07/discovered-our-parents-were-russian-spies-tim-alex-foley.

63 the real Gordon Lonsdale: For more see Gordon Corera, *The Illegal*, Amazon ebook, 2018.

64 "Heathfield does not know Russian": Bezrukov and Vavilov 2019 TV interview, https://tvzvezda.ru/news/qhistory/content/201931106-ftYfe.html.

64 "forget it, get rid of it": "Новые русские сенсации": "Русские разведчики-нелегалы в Америке," NTV, April 24, 2019.

64 In August 1995, the family sold off the business: Timothy Vavilov's affidavit submitted in Canadian court case.

66 The illegals' mission was to: In the early eighties, the KGB had planned to expand to six illegal "residencies" (meaning agents or couples) in place. Each was supposed to recruit three or four sources in the White House, State Department, Pentagon, and also think tanks and universities—these included the Hudson Institute, Rand Corporation, Columbia University's School of International Relations, and Georgetown's Center for Strategic Studies. Moscow also wanted active recruitment of students at Columbia, New York, and Georgetown Universities.

67 "the Royal Canadian Scotch Stagger": https://www.nytimes.com/2010/07/01/us/01cambridge.html and https://www.theglobeandmail.com/news/world/at-harvard-alleged-spy-hosted-a-scotch-drink-up/article4323320/.

67 "The main task of an agent is constantly climb": "Новые русские сенсации": "Русские разведчики-нелегалы в Америке," NTV, April 24, 2019.

6: THE SOURCE

71 One KGB colleague from those days: https://www.newsru.com/russia/04may2011/poteev_character.html. Further details of his life and career from http://www.rosbalt.ru/moscow/2011/05/16/849215.html and

https://www.newsru.com/russia/27jun2011/poteev_prigovor.html and "Media: Russian Spies in the United States Were Betrayed by the Belarusian KGB Officer," November 17, 2010, https://charter97.org/en/news/2010/11/17/33872/; *The Mitrokhin Archive,* vol. 2 (London: Penguin, 2005); and Vladimir Kuzichin, *Inside the KGB* (London: Andre Deutsch, 1990).

71 When he arrived every morning: The description of the residency is drawn from *Comrade J* by Pete Earley (New York: Berkley, 2007).

74 When he visited Yasenevo: Early, *Comrade J,* pp. 155–57.

76 His former colleagues would claim Poteyev sold them out: http://www.rosbalt.ru/moscow/2011/05/16/849215.html. Other accounts in Russia are more far-fetched. There was a confused story about the break-in at the family apartment having somehow been staged by the CIA to seal his disillusionment. But he had already been recruited by then. Other details from Grani.ru website June 28, 2011, http://www.rosbalt.ru/moscow/2011/05/16/849215.html; "Foes Among Friends," *New Times,* May 23, 2011; and *Kommersant,* July 8, 2011.

79 Only three officers had access to the personal files: Yuriy Senatorov, "For Outstanding Services: Defector Sentenced to 25 Years' Imprisonment in Absentia," *Kommersant Online,* Wednesday, August 10, 2011.

79 In Madrid in the summer of 1996, a rugged first secretary: The most detailed account comes in Mark Urban, *The Skripal Files* (London: Macmillan, 2018). Also additional private information.

80 By the middle of the decade, MI6 had reduced by two-thirds: Intelligence and Security Committee report, 1995.

7: THE INVESTIGATION

83 not who he said he was: Details in this chapter are drawn from FBI complaint against the illegals and related court documents as well as interviews with FBI officials involved in the investigation.

84 Peru: The FBI does not name Peru as the country. Peruvian press confirmed Pelaez's entry to the country around this time and that of Lazaro for the later 2007 meeting.

85 a political refugee: Vicky Pelaez, "la combativa 'espía' que vino del Cusco," *El Mercurio,* July 4, 2010.

86 "She was a very passionate woman": "Spy Mystery: Was NY Columnist a Wife Betrayed?" Associated Press, July 10, 2010.

86 "I first admired him for his knowledge": Pelaez's own account was published in https://mundo.sputniknews.com/opinion/20110805149975896/.

86 The real Juan Lazaro was a toddler: Richard Boudreaux, "Busted Russian Spy Wants Old Life Back," *Wall Street Journal,* August 7, 2010, https://www.wsj.com/articles/SB10001424052748703309704575413600124475346.

86 which he received in 1979: Ibid.

87 Russian reports, whose accuracy is hard to judge, suggest his marriage to Pelaez was genuine: *Kommersant,* November 11, 2010.

88 He praised Hugo Chávez, the populist left-wing leader of Venezuela: https://www.nytimes.com/2010/06/30/nyregion/30suspects.html.

90 This is a classic decades-old communications technique: All the illegals were trained on shortwave radio. It would later be superseded by newer techniques for regular communication, but they were expecting to be able to use it as a backup and primarily to receive rather than send information.

8: BREAKING AND ENTERING

91 It was the middle of the night: Information in this chapter drawn from interviews with FBI officials involved in the case and other documents associated with the case against the illegals.

92 great-granddaughter of former Soviet leader Nikita Khrushchev: https://www.bbc.co.uk/news/world-europe-10665123 and https://foreignpolicy.com/2010/07/14/the-spy-who-came-in-by-amtrak/.

93 The two families would barbecue together: "We can't believe our friends are Russian spies," *Daily News,* June 30, 2010.

96 "What I think of is these two eating a hamburger": Ricci speaking in an interview at https://www.usanetwork.com/insidethefbinewyork/blog/the-true-story-of-husband-and-wife-russian-spies-living-in-america.

97 "We owned almost every facet of their life": Robert Anderson quoted in *Fox Files,* March 25, 2013—"Operation Ghost Stories."

104 Tony Rogers of the Boston FBI field office: "Operation Ghost Stories: The Spies Next Door," *CNN Declassified,* July 22, 2017.

105 "The only goal and task of our Service": evidence presented in July 1, 2010, detention request to Judge Ellis.

9: PUTIN'S SPY FEVER

107 On December 20, 1999, hundreds of Russia's spies took refuge: Putin made good on his promise to FSB; *Moscow Times,* February 8, 2008.

109 As Litvinenko entered Putin's office in the Lubyanka: Litvinenko described his meeting with Putin in *The Gang from the Lubyanka*, quoted in the Litvinenko Inquiry report and also in his police interviews on his deathbed.

112 A few weeks after Putin became prime minister: Sutyagin case details from https://www.birminghammail.co.uk/news/local-news/revealed

-how-russian-spy-gave-248366 and "Revealed: The Incredible Story behind Midland House at Centre of Russian Espionage Scandal," *Sunday Mercury,* March 4, 2001.

112 US-Canada Institute: This was not part of the government but during the Cold War the institute had been an occasional target for espionage—with the Americans running at least two agents inside and the KGB having its own undercover operatives to keep watch. See Grimes and Vertefeuille, *Circle of Treason,* and Cherkashin's memoir.

112 "He was small, slight and mild-mannered": Spies may have used my seminar as cover. *Gloucestershire Echo,* March 1, 2001.

112 Sutyagin was arrested: Sutyagin's arrest and trial details from "Russia's Spy Mania: A Study of the Case of Igor Sutiagin," Human Rights Watch report, and Sarah Karush, "FSB Keeps 'Traitorous' Scholar Jailed," *Moscow Times,* February 16, 2000; "Sutyagin Verdict Worries Scientists," *Moscow Times,* April 7, 2004; "Ruing Exile, Russian Says He's No Spy," *New York Times,* August 13, 2010; reports by Ekho Moskvy Radio, April 7, 2004, and October 25, 2004.

113 What was Alternative Futures?: https://www.rferl.org/a/Interview_Igor_Sutyagin_Discusses_Spy_Swap_Life_In_England/2127860.html. Around 10,000 Russian scientists were reckoned to moonlight for foreign companies to make ends meet. "Sutyagin Verdict Worries Scientists," *Moscow Times,* April 7, 2004.

113 "see a young man destroyed": "Russian Faces Spy Trial Over Trip to UK," *Guardian,* February 26, 2001.

113 Sutyagin always said: "Ruing Exile, Russian Says He's No Spy," *New York Times,* August 13, 2010, and "Russian Pundit Dismissed Spy-Swap Analyst's Protestations as Lies," *Yezhednevny Zhurnal* website, August 26, 2010.

114 a high-ranking Russian spy spoke anonymously to the media: "Spy Case Shows That Russia Is Recovering," *RIA Novosti,* April 9, 2004.

114 "How else can you make dough like that?": "Spy Swapped in Deal with Russia Could Return to House in Maryland Suburb," *Washington Post,* July 11, 2010, and details from individuals with knowledge of the case.

116 He had been able to search FBI computer systems: Michael Rochford speaking on the podcast *Case File Review* with Jerri Williams, August 20, 2016.

116 His home was searched: Wise, *The Seven-Million-Dollar Spy.*

117 The KGB had never liked running an agent whose identity they did not know: Cherkashin, *Spy Handler,* p. 240.

117 $7 million payoff: Wise, *The Seven-Million-Dollar Spy,* and Gus Russo and Eric Dezenhall, *Best of Enemies* (New York: Twelve, 2018).

118 sat across a table from Zhomov: Milt Bearden, "No Letup in Search for 'the 4th Man,' " *Los Angeles Times,* June 15, 2003.

118 He was interrogated personally by Zhomov: Russo and Dezenhall, *Best of Enemies*, p. 244, and Ronald Kessler, *The Secrets of the FBI* (New York: Broadway, 2011).

10: TARGETING

119 Information in this chapter is drawn from interviews with FBI officials and others involved in the case or individuals who knew the illegals, as well as the indictment and other legal documents.

119 Colleagues remember a well-organized, hard worker: Friscolanti, "The Russian Spies Who Raised Us."

120 "He wanted to collaborate": Interview with Bill Halal, http://billhalal.com/?page_id=2 and https://www.washingtonian.com/2010/07/14/the-consultant-was-a-spy/.

120 "It was a piece of junk": http://archive.boston.com/news/local/massachusetts/articles/2010/07/01/business_opened_doors_for_spy_suspect/?page=full.

121 Heathfield tried to market his software to government agencies: https://www.bcdemocratonline.com/article/20100629/NEWS/306299963.

121 He met an employee of STRATFOR: https://worldview.stratfor.com/article/russian-spies-and-strategic-intelligence.

125 "We believe the SVR illegals may well have hoped to do the same thing here": Quoted in https://www.fbi.gov/news/stories/operation-ghost-stories-inside-the-russian-spy-case.

125 a company called Redfin, based in Somerville: https://www.nytimes.com/2010/07/01/us/01cambridge.html.

125 "She was nice, friendly, very normal": "Alleged Spies Always Strived for Connections," *Boston Globe*, June 30, 2010.

126 Timothy had been born in June 1990: Affidavit of Timothy Vavilov.

126 "We were trying to avoid making them typical Americans": "Новые русские сенсации": "Русские разведчики-нелегалы в Америке," April 24, 2019, https://www.youtube.com/watch?v=mDU879mQjSA&feature=player_embedded.

126 "then at least respect": Bezrukov interview, "Why Spies Are Like Scientists," *Russian Reporter*, October 11, 2012, https://expert.ru/russian_reporter/2012/40/pochemu-shpionyi-pohozhi-na-uchenyih/.

126 "As a family we loved to travel": Affidavit of Alexander Vavilov (aka Alexander Foley), March 2014.

128 "The highest class of intelligence": Bezrukov interview, "Why Spies Are Like Scientists."

128 He was a computer technician: "Russian spy ring aimed to make children agents," *Wall Street Journal*, July 31, 2012.

130 "we would hear him watching *The Sopranos*": "Operation Ghost Stories: The Spies Next Door," *CNN Declassified*, July 22, 2017.

131 SVR officers would say: Andrei Soldatov says he was told of other cases by SVR staffers in an interview on the Svobodnaya website, October 24, 2011.

11: ENTER ANNA

133 The young Russian was hard to miss: Details of Anna Chapman's life are drawn from interviews with a dozen individuals who knew her in London. See "The Photo Album of Suspected Russian Spy Anna Chapman and Her Ex-Husband Alex," https://www.telegraph.co.uk/news/picturegalleries/worldnews/7868331/The-photo-album-of-suspected-Russian-spy-Anna-Chapman-and-her-ex-husband-Alex.html?image=1 and https://www.telegraph.co.uk/news/worldnews/europe/russia/7866824/Russia-spy-Anna-Chapmans-husband-I-thought-I-knew-her.html.

133 This would have been a disappointment to Marcus Read: https://www.mirror.co.uk/news/uk-news/my-crazy-six-weeks-of-passion-with-russian-237500.

134 Ana later denied that there were any "secret motives": https://web.archive.org/web/20110713163907/http://www.thesun.co.uk/sol/homepage/features/3530430/Russian-spy-on-love-for-Brit-hubby.html.

135 When CIA director George Tenet went to talk to the leadership of Russia House: John Sipher and Steve Hall in the Cipher Brief special report "Foreign Influence, Domestic Division: The Kremlin and the American President."

137 Directorate S told agents to look for people who might have access to Western DNA databases: Kouzminov, *Biological Espionage*, p. 125.

139 transferred money from the United Kingdom to Zimbabwe: http://www.dailymail.co.uk/news/article-1295606/Redhead-spy-linked-UK-money-smuggling-ring.html.

139 "Anna had seductive charm": Richard Woods, "The Spy Next Door," *Sunday Times*, July 4, 2010.

140 100 billion pounds over the last twenty years: "Moscow's Gold," Foreign Affairs Select Committee report, May 21, 2018.

140 "significant risks" were being taken: UK Intelligence and Security Committee, 2003–2004 annual report.

140 with more aggressive spying abroad: "Russia Steps Up Espionage," *Jane's Intelligence Digest*, November 27, 2002.

141 By 2005, MI5 was warning: *Jane's Intelligence Review*, February 16, 2006.

142 "My dream was to go to London and find a rich husband": https://www.dailymail.co.uk/news/article-1315056/Agent-Anna-The-Man-Hunter-London-flatmate-reveals-Russian-spy-used-sex-prey-string-oligarchs.html.

142 specifically attending Boujis nightclub to meet Princes William and Harry: https://www.mirror.co.uk/news/uk-news/russian-spy-anna-chapman-had-wills-233061 and interview with Nicholas Camilleri.

146 target a ballerina in Paris: Pavel Sudaplatov, *Special Tasks* (London: Little, Brown, 1994), p. 247.

146 "sleeping agents" who would marry influential foreigners: Kouzminov, *Biological Espionage*, p. 104.

12: THE SPECTRE

149 no one would believe him: Details drawn from evidence to Litvinenko inquiry, particularly transcripts of police interviews. Also reporting carried out at the time and interviews with Marina Litvinenko and others who knew Alexander Litvinenko.

151 "Would you tell us who that person is?" asked the police officer: Scotland Yard record of November 20, 2006, interview with Edwin Carter (Litvinenko's alias).

155 Berezovsky in London: I found myself entering a strange world when I arrived at Berezovsky's swanky office in Mayfair, full of bodyguards and aides, before being taken into a large conference room where I sat alone with the former oligarch and one interpreter who helped him when he struggled to find an English word to convey his message that Putin was a murderer.

159 Litvinenko turned to Yuri Shvets: Yuri Shvets witness statement, Litvinenko Inquiry.

160 A major in the Russian tax police claimed Litvinenko: www.kp.ru/daily/23930/69736, 2007 interview.

160 "It began to dawn on me, that all was not what it seemed": Andrei Lugovoi at a press conference June 1, 2007.

161 "When the British agents started to approach me": Interview with *El País* in 2008, quoted in Litvinenko Inquiry report.

162 "piss on their grave": https://www.nytimes.com/2018/09/06/world/europe/skripal-poison-russia-spy-spain.html.

13: MOSCOW RULES

173 according to an official Russian investigation: https://www.newsru.com/russia/27jun2011/poteev_prigovor.html; https://www.newsru.com/russia/28jun2011/poteev.html and "Prosecutor Proposes to Sentence Poteyev to 25 Years in Absentia," Interfax, June 20, 2011.

175 On January 25, 2003, Poteyev suffered: https://www.novayagazeta.ru/articles/2010/11/21/573-predatelya-znali-vse.

14: THE CONTROLLER

182 a Special Reserve officer: Details from Kouzminov, FBI complaint, and also Oleg Gordievsky, who says these officers were also known as the Nagayev Group after their first commander. His own brother was one—traveling across Europe, Africa, and Asia. In one case he had to contact and bring back another illegal who had gone mad in Sweden. Oleg Gordievsky, *Next Stop Execution* (London: Macmillan, 1995), p. 136.

183 The central identity this man used was Canadian: Some reports say this passport was obtained through the Canadian embassy in South Africa.

183 The FBI believes he first came to the United States for operational work as early as 1993: The FBI Wanted poster for Metsos lists him as being wanted for acting as an unregistered agent in the United States from 1993 to 2005.

183 belonged to a car wash: "Accused Spy Studied at Vermont Military College," Associated Press, June 29, 2010, and http://old.themoscowtimes.com/sitemap/free/2010/7/article/how-a-suspected-spy-eluded-capture-in-cyprus/409771.html.

186 They were married on June 5, 2005: Other pairings like the Murphys and Heathfield/Foley had been sent out as couples. It is possible that in the Seattle case they had always been intended to operate together but their legends were documented differently as an American and Canadian and they were ready at different times. It is also possible that there was a probationary period for their relationship to make sure they were compatible before they got married.

186 Both attended the University of Washington: http://www.bothell-reporter.com/news/alleged-russian-spies-received-business-degrees-from-the-university-of-washington-bothell/.

186 high-tech companies like Microsoft and Amazon: One of the classes the illegals took at the university required students to create a website and complete a scavenger hunt with one of the tasks being to find Bill Gates's home address. The Russian consulate in Seattle would be closed in 2018 after it was reckoned by some to be a major spy facility with officers using it as their base from which to look for dead-drop and signal sites in places like Oregon beaches. https://www.politico.com/magazine/story/2018/03/29/what-really-went-on-at-russias-seattle-consulate-217761.

15: MURPHY STEPS UP

191 Paul Hampel was about to board: Federal Court in the matter of the person alleging to be Paul William Hampel, http://www.4law.co.il/canspi.pdf; https://www.theglobeandmail.com/news/national/alleged

-spy-beat-passport-curbs/article18178241/; and "Jet Setter's Life Veiled by Secrecy," *National Post*, November 22, 2006. The Balkans as his target is mentioned in documents from Registrar of Citizenship in Canada sent to Alexander Vavilov, dated August 15, 2014.

194 "We all high-fived ourselves because they didn't spot us": "Operation Ghost Stories: The Spies Next Door," *CNN Declassified*, July 22, 2017.

197 Gerard had been to Moscow on vacation in 2005: https://www.belfasttelegraph.co.uk/news/us-spy-ring-probe-donegal-firefighter-quizzed-28545338.html.

16: ANNA TAKES MANHATTAN

201 "Relentless": details in this chapter drawn from one of a number of interviews with individuals who knew her in New York.

202 A $1 million loan from: http://www.dailymail.co.uk/news/article-5500693/British-former-husband-Russian-spy-Anna-Chapman-DEAD.html.

202 "All dreams may come true if you act on it": The website for a final one, NYCRentals.com, was bought for around $25,000 in June 2010. Edward Lucas, *Deception* (London: Bloomsbury, 2012), p. 169.

203 "Not vulgar, but very flirtatious": https://abcnews.go.com/Blotter/russian-spy-ring-anna-chapman-accused-regular-nyc/story?id=11044883.

210 "He definitely didn't seem to be hiding anything": https://www.nj.com/news/index.ssf/2010/06/russian_spy_network_included_s.html.

210 "He seemed a lot younger than 27": https://www.alternet.org/story/147941/i_was_a_landlord_to_a_russian_spy,_and_he_was_a_total_slob.

210 "filthy imperialist economy from within": https://www.wsj.com/articles/SB10001424052748703374104575337533081334118.

210 Semenko returned to Russia in late 2009: https://www.washingtonian.com/2010/08/24/my-landlord-was-a-russian-spy/.

210 "Russia/China" expert: https://www.huffingtonpost.com/steve-clemons/another-twist-in-the-russ_b_639652.html and http://washingtonnote.com/close_call_with/.

211 A wireless network was detected: FBI complaint against Semenko.

211 Semenko emailed the president of the AFPC: https://www.washingtontimes.com/news/2010/jun/30/my-spy-story/.

211 Alexey Karetnikov: http://www.newsweek.com/miscrosofts-russian-spy-was-greasy-foreign-and-loved-snickers-217422 and https://www.bloomberg.com/news/articles/2010-07-14/microsoft-says-12th-alleged-russian-spy-worked-at-its-redmond-headquarters.

17: CLOSING IN

213 "taken a class in Suburbia 101": https://www.telegraph.co.uk/news/worldnews/northamerica/usa/7871348/Richard-and-Cynthia-Murphy-suburbias-Spies-Next-Door.html.

213 "Look what she did with the hydrangeas": https://www.nytimes.com/2010/06/29/world/europe/29spy.html.

214 "I was just struck at how accomplished she was": https://www.nj.com/news/2010/06/accused_russian_spies_children_nightmare.html.

215 But the KGB caught him: Previous cases of illegals turning against the system from Andrew and Mitrokhin, *The Mitrokhin Archive*, pp. 260 and 339.

217 "three major ways [for Richard Murphy]": July 1, 2010, detention request to Judge Ellis.

220 Patricof: "FBI Watched, Then Acted as Russian Spy Moved Closer to Hillary Clinton," TheHill.com, October 22, 2017, http://thehill.com/policy/national-security/356630-fbi-watched-then-acted-as-russian-spy-moved-closer-to-hillary.

221 "In the end, some of this just comes down to what it always does in Washington": Ibid.

221 "Several were getting close": http://articles.latimes.com/2011/oct/31/nation/la-na-russian-spies-20111101.

222 Colleagues remember him as grumpy and unhappy: "Accused Russian Spies Sure Fooled People in the Seattle Area," *Seattle Times*, July 1, 2010, https://www.seattletimes.com/seattle-news/accused-russian-spies-sure-fooled-people-in-the-seattle-area/ and James Ross Gardner, "The Russian Spies Who Fooled Seattle," SeattleMet, October 30, 2017, https://www.seattlemet.com/articles/2017/10/30/the-russian-spies-who-fooled-seattle.

224 skiing with his family in Whistler: Timothy Vavilov affidavit.

224 "We were alarmed when in the end when their team won": Vaviolova Tomsk TV interview.

226 Rudolf Herrmann, an illegal: Rudolf Herrmann case from Barron, *KGB Today;* Andrew and Mitrokhin, *The Mitrokhin Archive;* and https://ceospeaks.wordpress.com/category/from-russia-with-love/.

226 But the father was confronted by the FBI: This led to him acting as a double agent for two years (although KGB files suggest he was trying to send warning signs to Moscow about what was happening, which Moscow Center failed to notice).

227 Friends recall him being ambitious: https://www.gwhatchet.com/2010/07/08/details-emerge-about-gw-student-professors-tied-to-couple-in-russian-spy-case/.

227 "whatever makes me money": http://www.washingtonpost.com/wp-dyn/content/article/2009/11/14/AR2009111402704.html??noredirect=on.

227 saluted Mother Russia: "Russian Spy Ring Aimed to Make Children Agents," *Wall Street Journal*, July 31, 2012.

227 "just as ridiculous as it sounds": https://www.theguardian.com/world/2016/may/07/discovered-our-parents-were-russian-spies-tim-alex-foley.

228 "It's logical to presume, and we suspect he knew": Richard DesLauriers comments from "Son of Spies May Have Known Secret," *Boston Globe*, August 25, 2010.

231 "Had they been allowed to continue": Mueller comments from BBC, "Modern Spies."

18: DECISION TIME

234 was surrounded: http://www.spiegel.de/international/europe/betrayer-and-betrayed-new-documents-reveal-truth-on-nato-s-most-damaging-spy-a-693315.html.

234 NATO officials said the damage was comparable: https://www.eesti.ca/portrait-of-a-well-placed-mole-estonian-life/article46769 and Edward Lucas, *Deception* (London: Bloomsbury, 2012).

236 "It was one of the longer years in my career": Craig Fair comments from "Operation Ghost Stories: The Spies Next Door," *CNN Declassified*, July 22, 2017.

236 now needed to get out of Russia "immediately": Reference to "immediately" from John Brennan briefing Robert Gates. See Gates's memoir, *Duty* (New York: Vintage, 2015), p. 409.

237 "There were a number of reasons there": Robert Mueller comments from *Modern Spies*, BBC TV, April 2, 2012.

240 "As I looked round the White House": McFaul comments are from his book *From Cold War to Hot Peace* (London: Penguin 2018), pp. 202–4.

242 the whole arrest plan should be shelved: Account based on conversation with a number of officials involved in decision making and Robert Gates's memoir. Through spokespeople, Tom Donilon and Vice President Biden declined to comment.

242 Gates would understand: Gates talking to Panetta, in Gates, *Duty*, p. 409.

243 "This was part of a long story": Interview with Leon Panetta.

244 "That would be long, exhaustive": Interview with Leon Panetta.

244 to "minimize the fallout": Leon Panetta, *Worthy Fights* (New York: Penguin, 2015).

19: ESCAPE

246 making his final preparations to flee: A month or so earlier Poteyev had made a request to visit a mistress in Odessa for a week, saying she had just given birth to a child. But this was turned down, according to Russian reports.

246 One account says he rushed from a meeting: "Foes Among the Friends," *New Times,* May 23, 2011.

248 The FBI had a plan: Details drawn from interviews with FBI officers and FBI complaint against Chapman and Semenko.

253 Only later the Russians would realize: Russian reporting of the Poteyev trial and verdict.

255 "We believe that she was quickly processing": Todd A. Shelton of the Bureau, in "Operation Ghost Stories: The Spies Next Door," *CNN Declassified,* July 22, 2017.

257 Alexander Poteyev was on the move: Details of Poteyev's escape, primarily from Russian sources, come from "Poteyev Sentenced to 25 Years" and report by Yuriy Senatorov, "For Outstanding Services: Defector Sentenced to 25 Years' Imprisonment in Absentia," *Kommersant Online,* August 10, 2011.

260 "Once we knew the source was safe": Interview with Leon Panetta.

20: THE DAY IT ENDS

263 At the Heathfields' house: Details drawn from interviews by Heathfield and Foley with Russian media, interviews with FBI officials, and testimony of their children and other witnesses.

264 The children were left with about three hundred dollars each: Timothy Vavilov affidavit.

264 taken to an office in Boston: Details of Vavilova and Bezrukov arrest and initial detention from various interviews on TV, including https://www.youtube.com/watch?v=mDU879mQjSA&feature=player_embedded and "Why Spies Are Like Scientists," *Russian Reporter,* October 11, 2012, https://expert.ru/russian_reporter/2012/40/pochemu-shpionyi-pohozhi-na-uchenyih/ and https://www.rbth.com/articles/2012/10/18/russian_spy_reveals_his_secrets_19249.html; also Alex Foley's affidavit of March 2014, https://www.theguardian.com/world/2016/may/07/discovered-our-parents-were-russian-spies-tim-alex-foley.

267 The two daughters were later taken away: http://articles.latimes.com/2010/jun/29/nation/la-na-russian-agents-lives-20100630 and https://www.nytimes.com/2010/06/30/nyregion/30suspects.html.

268 "The hydrangeas did nothing wrong": Richard Woods, "The Spy Next Door," *The Sunday Times,* July 4, 2010.

269 Semenko was arrested: https://www.wsj.com/articles/SB10001424052748 703374104575337533081334118.

270 an FBI car intercepted them: Lazaro arrest scene, https://www.bbc.com /mundo/internacional/2010/06/100630_0427_espia_peruana_ao.

270 "Dozens and dozens of heavily armed FBI men": https://mundo.sputnik news.com/opinion/201108051499758961/.

270 "They knew my nickname": https://www.nytimes.com/2010/06/30 /nyregion/30suspects.html.

270 Alan Patricof wanted to give Hillary Clinton: http://www.dailymail .co.uk/news/article-5009373/FBI-surprised-Bill-Clinton-took-500-000 -Russia.html.

271 he was stopped by Cypriot police: Details of Metsos arrest, http://old .themoscowtimes.com/sitemap/free/2010/7/article/how-a-suspected -spy-eluded-capture-in-cyprus/409771.html and https://www.wsj.com /articles/SB10001424052748704699604575342991962704142.

272 Cyprus authorities went through the motions: Cypriot leaders' relationship with Moscow, https://www.theguardian.com/world/2010/jul/02 /russian-spy-ring-christopher-metsos and *Cyprus Mail* website, May 3, 2016.

272 SVR could have whisked him away: https://www.russiamatters.org /analysis/gangster-geopolitics-kremlins-use-criminals-assets-abroad.

21: THE SQUEEZE

276 "The old Cold Warriors in the CIA's clandestine service": Panetta, *Worthy Fights*.

276 What did Panetta think: Panetta and Fradkov conversation drawn from ibid.

279 "We had ten people plus their families": Interview with Leon Panetta.

281 He refused to give his real name: July 1, 2010, US Attorney for the Southern District of New York letter to Judge Ellis requesting detention order.

281 criminal complaint had to be carefully drafted: There were three different types of illegal: what the FBI called Cadre Illegals, like the Murphys; the Special Agent or True Name Illegals, like Chapman; and Semenko and the Traveling Illegal Metsos. Because they were not all connected to each other, they had to each be connected to Moscow Center either through their covert communications and tradecraft or through the role of Metsos and Russian embassy officers in a "hub and spoke" conspiracy. Particular focus was made on the points at which there were meetings—such as Metsos and Murphy or Murphy with Zottoli. The idea was to create a story that looped everyone in together.

282 "When I read the paper and saw lots of people's names": Vavilova Tomsk TV interview and https://www.kp.ru/daily/26986.4/4045797/.

282 "It's a circus. This is pure psychological pressure": "Feds: Alleged NY Spy Guy More Devoted to Russia than Son," *Gothamist,* July 2, 2010.

282 The charges were "ridiculous": http://articles.latimes.com/2010/jun/29/nation/la-na-russian-agents-lives-20100630.

283 Vavilova recalled: Heathfield and Foley recollections from Elena Vavilova Tomsk TV interview and thirty years undercover, Прямой эфир, July 9, 2019, Russian TV show.

284 he believed she would not want to go to Russia: "Swap of Accused Spies in Works?" *Seattle Times,* July 8, 2010.

284 Lazaro was pressured by a heavy-handed Russian: http://archive.boston.com/news/nation/washington/articles/2010/07/10/in_spy_swap_agents_were_pawns_in_a_practiced_game/.

284 "What's your name? Your real name": Richard Boudreaux, "Busted Russian Spy Wants Old Life Back," *Wall Street Journal,* August 7, 2010, https://www.wsj.com/articles/SB10001424052748703309704575413600124475346.

286 Igor Sutyagin was hard at work: https://www.rferl.org/a/Interview_Igor_Sutyagin_Discusses_Spy_Swap_Life_In_England/2127860.html; "Ruing Exile, Russian Says He's No Spy," *New York Times,* August 13, 2010; and https://web.archive.org/web/20100819011114/http://www.novayagazeta.ru:80/data/2010/087/11.html.

287 Sutyagin was taken to the office of the head of the prison: https://www.rferl.org/a/Interview_Igor_Sutyagin_Discusses_Spy_Swap_Life_In_England/2127860.html.

287 "It's a very simple deal": Sutyagin's recollections from "Ruing Exile, Russian Says He's No Spy," *New York Times,* August 13, 2010; https://www.rferl.org/a/Interview_Igor_Sutyagin_Discusses_Spy_Swap_Life_In_England/2127860.html; and Katherine Butler, "Spy Swap Scientist Pays a Price for His Freedom, *Independent,* September 29, 2010.

288 "I was between a rock and a hard place": "Ruing Exile, Russian Says He's No Spy," *New York Times,* August 13, 2010.

288 Some Russian officials: *Kommersant,* November 11, 2010.

288 The agreement was that: https://www.kp.ru/daily/24603/775004/.

288 the prisoners were expecting to be freed: Account from Russo and Dezenhall, *Best of Enemies.*

288 "you get so accustomed": Vavilova Tomsk TV interview.

289 did not want to go to Russia: Panetta, *Worthy Fights.*

289 "The only Russian thing my mother likes is vodka with passion fruit": "Sons Still Don't Believe Parents Were Spies," *New York Post,* July 9, 2010.

289 Alex and Timothy, Heathfield and Foley's two sons: Details of the family's recollections drawn from Alex Foley and Timothy Vavilov's affidavits and https://www.theguardian.com/world/2016/may/07/discovered-our-parents-were-russian-spies-tim-alex-foley; also Vavilov and Bezrukov TV interview, https://www.youtube.com/watch?v=6CvRgpSFxSA.

290 "I had mixed emotions in that cell": Vicky Pelaez 2011 interview with *Revista Caretas*, https://nuestragente2010.wordpress.com/%E2%96%BA-vicky-pelaez-regresa-al-peru/.

291 Her lawyer afterward said she was expecting: "The Russian Swap-Stakes—Anna & Spies Traded to the Reds," *New York Post*, July 9, 2010.

22: VIENNA

293 "We've got to stop meeting like this": Russo and Dezenhall, *Best of Enemies*, p. 287.

293 Sutyagin would later say: https://www.kp.ru/daily/24603/775004/.

293 "Do you think it's a trick?": Recollections of flight from Moscow to Vienna, https://web.archive.org/web/20100819011114/http://www.novayagazeta.ru:80/data/2010/087/11.html, and Russo and Dezenhall, *Best of Enemies*.

297 Vavilova later recalled: Email from Vavilova and https://www.kp.ru/daily/26986.4/4045797/.

297 Sutyagin was given instructions as he disembarked: "Ruing Exile, Russian Says He's No Spy," *New York Times*, August 13, 2010.

297 "It took 40 seconds": *Modern Spies*, BBC TV.

297 "Nice try, motherfucker": Russo and Dezenhall, *Best of Enemies*, p. 289.

298 no band or big welcoming party: Sutyagin's arrival in the UK, https://web.archive.org/web/20100819011114/http://www.novayagazeta.ru:80/data/2010/087/11.html.

299 he dreamed of returning to Russia: https://republic.ru/russia/v_tyurme_ya_rasskazyval_zaklyuchennym_o_kvantovoy-431251.xhtml.

299 warnings from multiple people: "Freed Russian Analyst Sutyagin Yearns for Home," Associated Press, August 13, 2010, and https://www.rferl.org/a/Interview_Igor_Sutyagin_Discusses_Spy_Swap_Life_In_England/2127860.html.

23: ANGER

302 the illegals were taken to the forest at Yasenevo: https://www.mk.ru/politics/article/2010/07/12/515908-razvedchikam-dali-srok-na-rodine.html.

303 The series was one of Putin's favorites: Gaddy and Hill, *Mr. Putin*, p. 344, and https://www.theguardian.com/world/2010/jul/25/vladimir-putin-russian-spy-ring.

304 He had died at age fifty-three: Tretyakov's death, https://web.archive.org/web/20100731201345/https://www.novayagazeta.ru/data/2010/081/13.html.

305 "died of a heart attack": Tretyakov's death, http://www.peteearley.com/2010/07/09/sergei-tretyakov-comrade-j-has-died/.

305 An autopsy confirmed the cause of death: http://www.nwfdailynews.com/news/piece-33102-chokes-russian.html.

306 This was a full-blown crisis: *Kommersant*, November 11, 2010.

306 The FSB had already been given the power to conduct operations overseas: "Politics": "The Chekists Against the Spies," *Novaya Gazeta*, November 15, 2010.

307 the cost of training and supporting illegals: https://www.newsru.com/russia/28jun2011/poteev.html.

307 "What damn illegals?": Soldatov comments, https://themoscowtimes.com/articles/why-a-young-american-wants-to-be-a-russian-spy-16845.

308 Old SVR hands did their best: "Former Russian Intelligence Officers Analyze Recent Spy Scandal," *Novaya Gazeta*, November 26, 2011.

308 "we talked deep into the night": Friscolanti, "The Russian Spies Who Raised Us."

308 "In the end, they had an understanding": Bezrukov interview, "Why Spies Are Like Scientists," *Russian Reporter*, October 11, 2012, https://expert.ru/russian_reporter/2012/40/pochemu-shpionyi-pohozhi-na-uchenyih/.

308 "They understood us": Vavilova Tomsk TV interview.

309 take down Morse code at twenty groups a minute: "30 Years Undercover," Прямой эфир, July 9, 2019.

309 taken on a tour: Trips around Russia described in https://www.theguardian.com/world/2016/may/07/discovered-our-parents-were-russian-spies-tim-alex-foley.

309 feared he might have been turned: Gordievsky, *Next Stop Execution*, p. 142.

309 "He was sore as hell": "The Illegals," *Inside Story*, BBC, 1992.

310 Michael Zottoli: Interfax December 24, 2012, and https://fief.ru/img/files/180707_list_RUS_EU_participants.pdf.

310 "Class with Andrei Bezrukov": http://www.thesetonian.com/2018/04/11/student-journeys-to-moscow-for-foreign-policy-studies/.

310 "Just watch *The Americans*": Heathfield talking to Russian students in 2018, https://www.youtube.com/watch?v=u_rX3_A6am8.

310 "The result is quite close to reality": Heathfield on *The Americans*, https://www.kommersant.ru/doc/2781713.

311 spoke together mainly in English: Heathfield on language, https://www.kommersant.ru/doc/2781713 and http://www.tsu.ru/podrobnosti/andrey-bezrukov-u-razvedchika-na-dushe-koshki-ne-skrebut/.

311 "What I regret most of all": Bezrukov and Vavilova TV interview 2019, https://tvzvezda.ru/news/qhistory/content/201931106-ftYfe.html.

312 "There are certainly no regrets": Vavilova Tomsk TV interview.

313 "I feel I have lost touch with my previous self": Alex and Timothy Vavilov comments from their Canadian court affidavits.

314 "We don't have a very close relationship": Bezrukov and Vavilova TV interview, 2019, https://tvzvezda.ru/news/qhistory/content/201931106-ftYfe.html.

315 "He doesn't want to stay in Russia": Boudreaux, "Busted Russian Spy Wants Old Life Back."

315 "very strong words": Vicky Palaez 2011 interview with *Revista Caretas*, https://nuestragente2010.wordpress.com/%E2%96%BA-vicky-pelaez-regresa-al-peru/.

315 "collateral damage": https://mundo.sputniknews.com/opinion/20110805149975896/.

316 "sleazy glitz": Lucas, *Deception*, pp. 14 and 190.

316 Anna Chapman and Vladimir Putin: One person even claimed that Chapman had said she had spent time with the other illegals at a villa on the Black Sea with Putin and that she had been singled out for a ride with him in his personal submarine. https://www.politico.com/states/new-york/albany/story/2012/01/the-big-russian-life-of-anna-chapman-ex-spy-067223.

317 "She quickly became disappointed in Alex": https://www.telegraph.co.uk/news/worldnews/europe/russia/7873299/Friends-of-Russian-spy-Anna-Chapman-call-ex-husband-a-rag.html.

317 "We didn't much like Poteyev himself as a person": Vavilova Tomsk TV interview.

317 "He had a choice. He made his choice": Bezrukov interview, "Why Spies Are Like Scientists," *Russian Reporter,* October 11, 2012, https://expert.ru/russian_reporter/2012/40/pochemu-shpionyi-pohozhi-na-uchenyih/.

318 "It was no coincidence": Fradkov comments, interview with *Rossiyiskaya Gazeta,* December 20, 2010.

318 "We know who and where he is": *Kommersant,* November 11, 2010.

318 of his regular TV appearances: Transcript of Russian prime minister Vladimir Putin's annual Q&A session, broadcast on December 16 by Rossiya 1 and Rossiya 24 TV channels and Mayak and Vesti FM radio stations and reproduced in English on Putin's official website.

24: STILL AMONG US

322 Frith was another illegal: Details of the Frith case from Pierre Briancon, "The Story of a Russian Illegal," *Politico,* June 17, 2016, and *New Zealand Herald,* March 14, 2018. A Russian spy operated under a fake New Zealand identity and additional information from sources with knowledge of the case.

323 A Russian report after Poteyev's trial: *New Times* website, May 23, 2011.

323 the EU and NATO uncovered twenty illegals: NATO figures of illegals from https://twitter.com/RidT/status/1022253545800650752/photo/1.

324 The case was a perfect example: Details of Anschlag case from *Welt am Sonntag* website on July 19, 2015.

326 the world's largest video platform: Use of YouTube, https://www.welt.de/politik/deutschland/article112870150/Die-Vorliebe-von-Alpenkuh1-fuer-Cristiano-Ronaldo.htm.

326 The Anschlags served relatively short prison sentences: Details of Anschlag case from *Focus* magazine, June 25, 2012.

327 The couple moved to Belgium: "Belgian Spy Couple," http://www.mo.be/en/news/belgium-chases-russian-spy-couple and a detailed report in *Le Monde,* July 25, 2018.

327 A few weeks later a sensational headline: "MI5 Quizzed MP's Aide on 'Spy Links,' *Sunday Times,* August 29, 2010. The details of MI5's claims and the evidence presented in the legal battle are drawn from Special Immigrations Appeals Commission transcripts.

330 Evgeny Buryakov arrived in New York: Details of Buryakov case drawn from indictment, DOJ press release on his conviction, and https://www.bloomberg.com/news/articles/2016-11-15/the-spy-who-added-me-on-linkedin.

335 Page says he has always sought to: Letter to House Permanent Select Committee on Intelligence from Carter Page, May 22, 2017.

336 The FBI never believed that Page: The 2008 contact and further detail on Page are cited in the Mueller report.

336 "This isn't about just stealing classified": Ricci quoted in information. https://www.bloomberg.com/news/articles/2016-11-15/the-spy-who-added-me-on-linkedin.

25: A NEW CONFLICT

337 the annual meeting of the Valdai Group: Details of the Valdai dinner from participants and https://www.brookings.edu/on-the-record/putins-next-move-in-russia-observations-from-the-8th-annual-valdai-international-discussion-club/; https://nationalinterest.org/commentary/cold-autumn-russia-the-valdai-club-2011-6172; and also conversations with other attendees.

340 Putin is said to enjoy watching *House of Cards:* Mikhail Zygar, *All the Kremlin's Men* (New York: PublicAffairs, 2017), p. 271.

341 an FSB officer pulled the baseball cap: https://www.youtube.com/watch?v=Je5cjpDwAbU&feature=player_embedded.

341 liaison discussions about counterterrorism: A bombing at the Boston Marathon the previous month had led to attempts to improve contacts after it emerged Russian authorities had notified the United States about the extremist views of one of the attackers two years earlier. American in-

telligence officials involved at the time dismiss the idea that the Russians were helpful, saying they received those details along with four hundred other names in a dossier a foot thick and seeded with names of defectors to cause trouble for them by having them identified as extremists so they could not travel.

343 In February 2013, a military newspaper: https://foreignpolicy.com/2018/03/05/im-sorry-for-creating-the-gerasimov-doctrine/.

344 although the CIA may have been able to get: It was reported that this source was extracted in 2017 amid concerns he might be identified. They are alleged to have reported on US election interference in 2016 but were operating for a number of years.

26: THE NEW ILLEGALS

353 What the man building the cage did not know: https://money.cnn.com/2018/02/20/media/internet-research-agency-unwitting-trump-supporters/index.html.

354 the KGB had recruited a Democratic Party activist: Mitrokhin archive, cited by US Intelligence Community assessment.

354 The Ghost Stories illegals were tasked: This is referred to in the US 2016 intelligence community assessment of Russian interference in the election.

354 It was registered in July 2013: https://comprop.oii.ox.ac.uk/research/ira-political-polarization/. The Oxford study sees the first activity in 2009 inside Russia and low-level English language work from 2013 but picking up significantly in 2014.

355 Project Lakhta's job: Details "Project Lakhta": indictment and other documents of Elena Khusyaynova by David Holt of FBI Washington Field Office, unsealed September 28, 2018.

358 "Our goal wasn't to turn the Americans toward Russia": https://tvrain.ru/teleshow/bremja_novostej/fabrika-447628/.

358 Young men and women: Adrian Chen's account in the *New York Times* gives the earliest insight into it: https://www.nytimes.com/2015/06/07/magazine/the-agency.html.

358 Twitter was where it began: Details of life within the IRA, https://tvrain.ru/teleshow/bremja_novostej/fabrika-447628/.

360 As early as 2014: https://www.politico.com/story/2017/08/14/obama-russia-election-interference-241547.

369 "We see that in the West, including the United States": http://iz.ru/news/623822.

369 a profile of Donald Trump: Bezrukov's profile of Trump is at http://valdaiclub.com/a/reports/report-donald-trump-a-professional-profile/.

27: THE NEW WAYS

371 Two days later, she asked Torshin: Kevin Helson affidavit, pretrial detention request from the FBI, and government's opposition to defendant's motion for bond review, https://www.courtlistener.com/recap/gov.uscourts.dcd.198600/gov.uscourts.dcd.198600.26.0_1.pdf.

371 "If I'm a spy": https://newrepublic.com/article/153036/maria-butina-profile-wasnt-russian-spy.

372 The CIA had reportedly believed: https://www.npr.org/2018/05/11/610206357/documents-reveal-how-russian-official-courted-conservatives-in-u-s-since-2009.

374 The FBI was adamant: This and other details from pretrial detention request are from the FBI.

374 Butina's lawyers said: Motion to dismiss pretrial detention.

375 the Russian intelligence services would be able to use it for years to come: Robert Anderson declaration in support of sentencing memorandum in the Butina case.

28: REVENGE

387 Department 8 of Directorate S: Andrew and Mitrokhin, *The Mitrokhin Archive*, pp. 505–10.

387 investigations have been so cursory: In Britain, investigations have been poorly handled, like that by Surrey Police into Russian exile Alexander Pereplichny, who dropped dead while jogging.

388 "Treason is the gravest crime possible": Interview with the *Financial Times*, June 28, 2019.

388 former top SVR officials had got in touch with European counterparts: "Skripal: A Strange Message from the SVR," *Intelligence Online*, April 11, 2018.

390 Russian facilities: used for communications monitoring. See "Russia Carried Out a 'Stunning' Breach of FBI Communications System, Escalating the Spy Game on U.S. Soil," Yahoo News, September 16, 2019.

391 "TDY" travelers: Private information and "TDY Spies Used to Track Defectors," Reuters, https://www.reuters.com/article/us-usa-russia-spies/fewer-russian-spies-in-u-s-but-getting-harder-to-track-idUSKBN1H40JW and https://edition.cnn.com/2018/04/25/politics/expelled-spies-russians-tracking-defectors/index.html.

392 "A person who has committed treason": *Kommersant*, July 8, 2016.

392 Poteyev, though, was not hard to find: BuzzFeed was able to find out considerable details of Poteyev's life in Florida. See, for example, https://www.buzzfeednews.com/article/alexcampbell/alexander-poteyev-cia-vladimir-putin-russian-spy-undercover.

393 even as he approached the home: Details of the approach to a defector's home were first reported by the *New York Times:* https://www.nytimes.com/2018/09/13/us/politics/russian-informants-cia-protection.html.

394 "We believe we eviscerated that [program]": Evanina comments, https://www.youtube.com/watch?reload=9&v=CtsLDjWXQAI.

395 a KGB officer back in Dresden: Putin referring to his work with illegals, *Times* of London, June 26, 2017.

396 began to appear on talk shows: "Bezrukov and Vavilova Talk Show Appearance—30 Years Undercover," Прямой эфир, July 9, 2019.

397 "I wish you good health": Putin visit to Yasenevo, speech released by the Office of the President of Russia, June 28, 2017.

398 "I am an average undercover agent": https://www.youtube.com/watch?v=6CvRgpSFxSA.

INDEX